SpringerBriefs in Applied Sciences and Technology

For further volumes:
http://www.springer.com/series/8884

Yan Chai Hum

Segmentation of Hand Bone
for Bone Age Assessment

 Springer

Yan Chai Hum
Universiti Teknologi Malaysia
Skudai-Johor
Malaysia

ISSN 2191-530X   ISSN 2191-5318   (electronic)
ISBN 978-981-4451-65-9   ISBN 978-981-4451-66-6   (eBook)
DOI 10.1007/978-981-4451-66-6
Springer Singapore Heidelberg New York Dordrecht London

Library of Congress Control Number: 2013936534

Printed on acid-free paper

Springer is part of Springer Science+Business Media (www.springer.com)

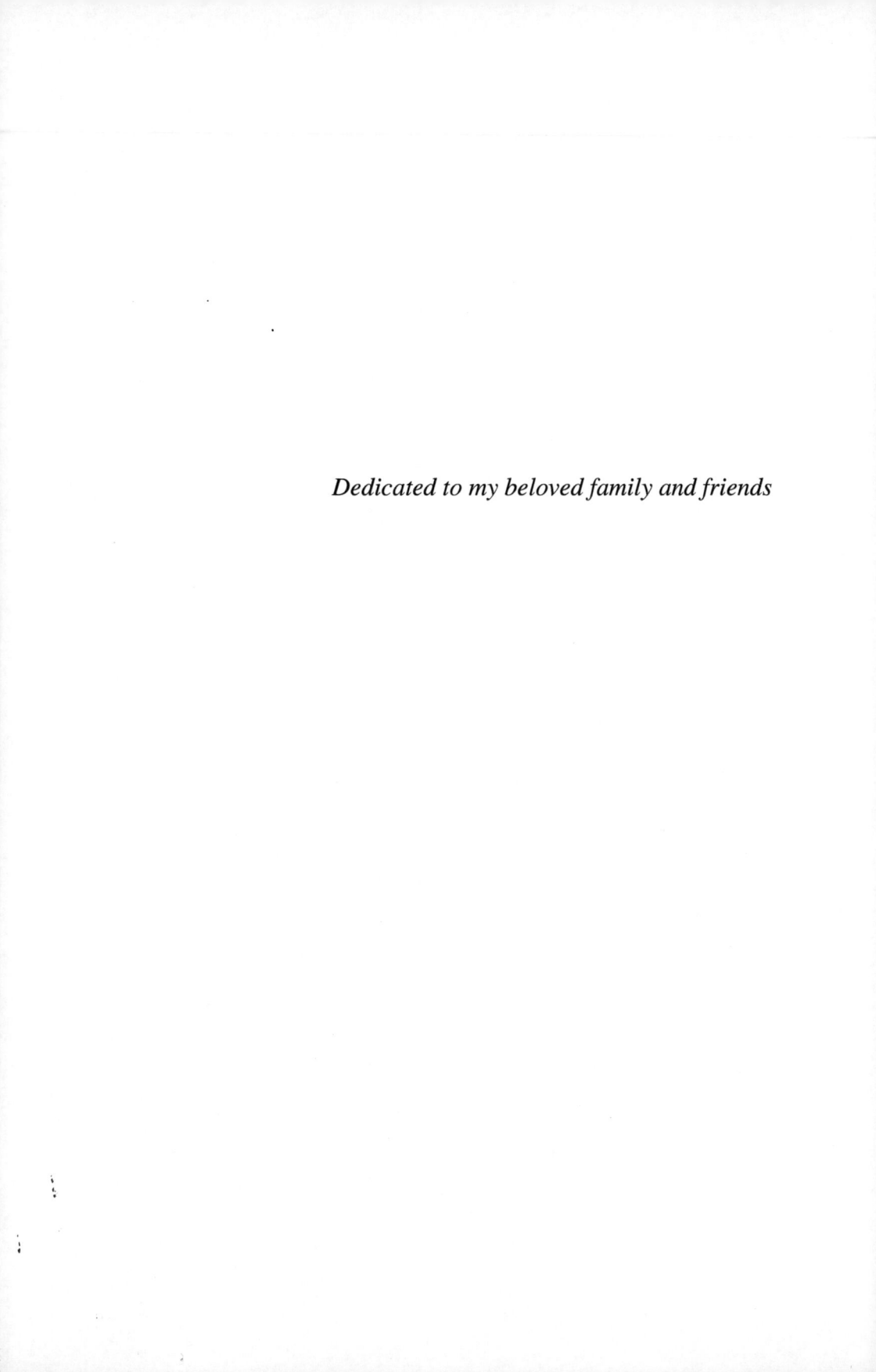

Dedicated to my beloved family and friends

Preface

Bone age assessment is an examination of ossification development with the purpose of deducing the skeletal age of children to monitor their skeletal development and predict their future adult height. Conventionally, it is performed by comparing left hand radiographs to standard atlas by visual inspection. This process is subjective and time consuming. Therefore, automated inspection system, namely, computer-aided skeletal age scoring system has been established to overcome the drawbacks. However, the computerized system invariably confronts with problem in hand bone segmentation, which is the most crucial procedure in the automated system that determines the success or failure of subsequent processes of the system. The current manual or semi-automated segmentation frameworks have impeded the system from becoming truly automated and efficient. The objective of this book is to provide a solution to large variability in pixel intensity, uneven illumination, complicated anatomical structure, and artifacts in automated bone age assessment system. The task is accomplished by first applying the modified histogram equalized module, then undergoing the proposed automated anisotropic diffusion, followed by a novel fuzzy quadruple division scheme to optimize the central segmentation algorithm, and finally the process ends with an additional quality assurance scheme. The designed segmentation framework works without demanding scarce resources such as training sets and skillful operator. The result analysis of the resultant images has shown that the designed framework is capable of separating the soft tissue and background from the hand bone with relatively high accuracy despite omitting the resources.

Acknowledgments

Foremost, I am so indebted and would like to thank heartedly my advisor, Dr. Tan Tian Swee. Under his effective guidance, I have built up my intellectual foundation on this interdisciplinary field and the proper attitude toward scientific research. Furthermore, thanks to his consistent motivations, patience, and enthusiasm especially in providing me freedom in pursuing my interested research topics. I am delighted to have such a helpful supervisor that always accessible and willing to share his knowledge.

Special thanks to Medical Implant Technology Group (MEDITEG), Center for Biomedical Engineering (CBE), and Universiti Teknologi Malaysia (UTM) for providing me various resources and stimulating research environment that are highly conducive to the pursuit of knowledge. Besides, I would like to express my sincere gratitude to all my research colleagues, friends and all the academics staffs, as well as the management staffs in my faculty. Their generosity to give valuable suggestion and constructive criticism to my works is sincerely appreciated.

Finally, heartfelt gratitude goes to my parents and my family members for all their supports and endless love in upbringing me. My future pursuit in research career would not have been possible without their encouragement, blessings, and inspiration. At this moment of accomplishment, they are those who deserve the credit. Words fail me to convey my eternal appreciation toward their sacrifices in various ways throughout my life thus far.

Contents

Abbreviations

2D	Two dimensional
3D	Three dimensional
ACR	Adaptive crossed reconstruction
AMBE	Average mean brightness error
ASD	Average structural different
ASM	Active Shape Model
BAA	Bone age assessment
BARNAE	Bounded-area Restoration and Non-bounded Area Elimination
BBHE	Bi-histogram equalization
BPS	Brightness preservation score
CAD	Computer-aided diagnosis
CDF	Cumulative density function
CER	Clustering efficacy ratio
CHLA	Children's hospital Los Angeles
DPS	Detail preservation score
DrDog	Derivative difference of Gaussian
DSIHE	Dualistic sub-image histogram equalization
FOC	Figure of certainty
FOM	Figure of merits
FRAG	Fragmentation
GHE	Global histogram equalization
GLCM	Grey Level Co-occurrence Matrix
GP	Greulich–Pyle
GVF	Gradient vector flow
HDT	Homogeneity discrepancy test
HE	Histogram equalization
MBOBHE	Multipurpose beta optimal bi-histogram equalization
MMBEBHE	Minimum mean brightness error bi-histogram equalization
MSE	Mean squared error
NBPS	Normalized brightness preserving score
NCL	Normalized cluster location
NDE	Normalized difference expectation

NDPS	Normalized Detail preservation score
NOCS	Normalized optimum contrast score
NRMS	Normalized root mean squared
OCS	Optimum contrast score
PDF	Probability density function
PMAD	Perona–Malik anisotropic diffusion
PSNR	Peak signal-to-noise ratio
RBD	Relative brightness difference
RCD	Relative contrast different
RMSHE	Recursive mean separate histogram equalization
ROI	Region of Interest
RSIHE	Recursive sub-image histogram equalization
RSWHE	Recursively separated and weighted histogram equalization
RUS	Radius, ulna and short bones
SRAD	Speckle reducing anisotropic diffusion
TW	Tanner-Whitehouse

Symbols

T_i	Threshold value
m	Mean of kernel centered
s	Standard deviation of kernel centered
$\lvert G(x,y) \rvert$	Absolute gradient magnitude
w	Kernel width
$N[(x,y)]$	Immediate neighboring pixels of each element
$S(I,J)$	Similarity measure
R_i	Region
m_i	Regional minima
$Bn(mi)$	Flood at stage n of m_i
S	Parametric domain
$v(.)$	Embedded parametric curve
$w_1 (.)$	Weight function that manipulates contour tension
$w_2 (.)$	Weight function that manipulates contour rigidity
$E_{int} [.]$	Internal energy of contour
$E_{ext} [v(s)]$	External energy of contour
$D(T)$	Diffused image at T-iteration
$V(T)$	Local variance of D(T)
μx	Mean of input image
μy	Mean of output image
$\beta(.)$	Beta function
σx	Standard deviation of input image
σy	Standard deviation of output image
$argmax(.)$	Argument of the maximum
$argmin(.)$	Argument of the minimum
$\vec{x}$	Variable vector
$\vec{f}(\vec{x})$	Vector objective function
X_L	Lower input sub-image
X_U	Upper input sub-image
$c_L(.)$	CDF of lower sub-image
$c_U(.)$	CDF of upper sub-image
$Y_L(.)$	Lower output sub-image

$Y_U(.)$	Upper output sub-image
n_L	Number of pixel in lower sub-image
$P_L(.)$	PDF of lower sub-image
$P_U(.)$	PDF of upper sub-image
$g(\|\|\nabla \mathbf{I}\|\|)$	Diffusion strength function
∇l	Image gradient
$div(\nabla l)$	Divergent operator
$c(q)$	Diffusion coefficient function of SRAD
$q_0(t)$	Coefficient of variation at the time, t
$C_j(T)$	Cluster center of jth cluster at Tth iteration
σ	Tolerant error value
$D_{ji}^{(T)}$	Euclidean distance between jth cluster and ith element at Tth iteration
$HOM(.)$	Homogenous function
$P_{(i,j)}$	Probability a group of spatial related pixel intensity

Chapter 1
Introduction

Abstract This chapter informs the readers about the background of bone age assessment of which include the clinical practicality, and the pros and cons of the conventional methods of this assessment. After that, we explain the automated system that is capable of substituting the conventional methods by further elaborating the pitfalls and constraints of this alternative. Then, we define our objective in this book and delimit the discussion scope of this book follow by stating the contributions of this book for the field in hand bone segmentation. Lastly, we provide the outline of the book.

1.1 Introduction

It is not trivial to have a distinct definition to physical maturity, not to mention an accurate quantitative measurement; conventional stature measurement, which does not assure common end points, can hardly be used to measure maturity. In other words, one is not certain about the maturity of a child by his or her chronological age; for instance, knowing that a child whose height is considered 'tall' compared to other children at same age does not indicate definitely that the child is more matured than the other children. Therefore, stature measurement is not suitable to measure maturity. However, there are some defined events that are definite to occur in normal individuals. Those events are suitable to measure the maturity. Events during puberty throughout adolescence such as eruption of a certain tooth, occurrence of first menstrual period, degree of testicular and appearance of pubic hair can be used as indicator for the maturity. It is deducible that an individual that has undergone a particular event is more matured than the other individuals that have not undergone the event. Not only limited to the occurrence, but the development of the events can further provide strong evidence to the degree of maturity such as the breast development for girls or the penile development for boys. Such events are invariably called as developmental 'milestones' to indicate how far an individual has travelled along the pathway to full maturity.

The utilization of events sequence to measure maturity is not without weaknesses. The main weakness is that the events are all not closely spaced [1]. The coarsely spaced event sequences lead to incomplete and uneven coverage of developmental age span [1]. This motivates the usage of hand and wrist bones which contain sufficient number of sequences that could cover the age span. Some bones are visible during the fetal life while some are visible during the certain ages of the individual [2]. All the bones eventually develop into constant final shape and the stages of the sequence in each bone are able to be recognized through radiographs. The invariant event sequences of total twenty bones: radius, ulna, metacarpals, phalanges and carpals develop in all stages along the pathway to full maturity. Thus, by inspecting and combining the evidences from the skeletal development of hand and wrist bones development, the maturity can be assessed and measured. This kind of inspection from hand bones development to deduce the bone age of the children is called Bone Age Assessment (BAA).

BAA or bone maturity assessment is applied clinically to gauge the skeletal development of children and adolescents [3]. Owing to the inefficacy to measure maturation age by chronological age, the skeletal maturity age is used to indicate the growth disorders and the to predict the future height [4]. Among all the skeletal bones, left hand has been proved to be a suitable indicator to gauge the skeletal maturation. Hence, the development of left hand skeletal is employed to represent the biological maturity depending on manifestations such as ossification area development and position of calcification [5]. Endocrine disorders [6], chromosomal disorders early sexual maturation, and others can be detected via the discrepancy between the skeletal age and biological age [7].

Generally, two popular types of assessment systems are currently adopted by pediatricians [8]: Greulich-Pyle atlas [9] method [10] and Tanner-Whitehouse (TW2) methods [11]. Physicians that adopt the Greulich-Pyle method, compare the left hand bone radiograph of patients to the standard atlas to deduce the skeletal age; TW2 method, on the contrary, is a index system via point collection. The TW2 system is then extended to TW3 system [1]; one of the differences between them is that TW2 collects the maturity evidences from twenty bones score consisting of the combination of Radius, Ulna and Short bones (RUS) and carpals. Instead, TW3 collects maturity evidences separately from RUS or Carpal scores on the ground that the combination of both has less accuracy in deducing the bone age [12].

Both methods, in terms of reliability, remains debatable in light of the assessment nature are based on visual inspection. Therefore, both methods are related to high dependency on the physician background knowledge and excessive time-consumption [13–16]. Thus, recently, many computer-aided systems of BAA have been established, particularly for TW2 method that is more convenient for computerization process [17–19]. However, most of the systems are still unstable attributable to the inadequate stability and inability to perform the BAA autonomously [20].

This research aims to establish a segmentation framework towards the generic segmentation development by presenting an alternative solution to current existing segmentation techniques so that it is able to identify the bones of the hand skeleton and outline their contours effectively when demanded resources for other sophisticated segmentation algorithms are scarce. In next sub-section, the discussion is on the background of the problem about the computerized system, the approaches of other researchers that attempted to address the problem, and the extent of the existing approaches in addressing the problem.

1.2 Background of the Problem

The current existing segmentation methods and frameworks either involve in threshold settings or are too dependent on certain resources and image features. This indicates that improvement on hand bone segmentation is necessary in order to practically realize the fully automated computer-aided skeletal age scoring system. Thus, this research is to explore this improvement aiming to establish an fully automated segmentation framework that is accurate yet remains less dependent on external resources.

1.3 Problem Statements

There are abundant of hand bone segmentation techniques found in the literatures, but very few of which, if ever, is functioning as an effective and yet remains fully automated. The research problem, therefore, is to explore the question: Is there any method that can realize the goal of performing the automated segmentation task that is relatively much more effective than fundamental segmentation techniques but yet is unaffected by constraints such as training sets and human intervention that are invariably pertain to sophisticated techniques?

The factors that associate with the problem are presented as follow:

1. The variability in hand radiographs deviates across different input sources and different age groups of the subjects in radiographs. This variability impairs the performance consistency or precision of segmentation technique or frameworks once the input radiographs are not as expected.
2. Devoid of prior knowledge of computational algorithm in recognizing pattern that can be easily perceived by human. As a consequence, most segmentation framework necessitates explicit labors and hence this problem violates automaticity.
3. The inherent bone intensity property in radiograph that stem primarily from the variations in anatomical density of different parts of the hand bone. As a consequence, two adverse properties for segmentation performance take place:
   a. The overlapping range of pixel intensity for the cancellous bones, the soft-tissue regions and the compact bones.
   b. The non-uniformity within the same category of bone such as cancellous bone or the cortical bone the radiograph intensity is not evenly distributed.

4. The uneven brightness intensity difference between the edge border of compact bone and soft-tissue regions and also between soft-tissue regions and background further complicates the problem.
5. The existing segmentation methods, as illustrated in Chap. 2, lacks of adequate considerations encompassing various aspects of being a comprehensive technique to perform the segmentation tasks or being too dependent on limited resources such as the availability of training hand bone samples over all age ranges, computational complexity and the knowledge background of operators.
6. Most of the existing works do not incorporate quality assurance process in the designed segmentation frameworks. As a consequence, the subsequent processing stage such as feature analysis for bone age assessment will accept inferior quality segmented hand bone as input and hence produces final result that is not reliable.

Partitioning hand bone from radiograph background and soft-tissue region is the first stage of computer-aided skeletal age scoring system; the performance of this stage underpins the success of subsequent procedures of bone age assessment which in turn, affects the final result. Despite being the significant step in computerized BAA, the automated segmentation remains a challenging problem owing to above mentioned problems. Conventional segmentation methods and the currently developed segmentation framework designed for hand bone segmentation are generally impractical to be implemented attributable to high dependency on various resources. The most problematic of which is the number of user-specified parameters in developing a fully automated segmentation without any human intervention.

1.4 Research Objectives

The main objective of this research is to propose, design and develop an enhanced framework of segmentation that is capable to automatically extract hand bone structures including the ossification sites without the soft-tissue region and radiograph background. To overcome the insufficiencies of the conventional segmentation techniques in challenging problem of hand bone segmentation attributable to large variability in pixel intensity, uneven illumination, complicated anatomical structure and several type of artifacts, several modules have been proposed or extended to meet the desired properties for the purpose of increasing the applicability of hand bone segmentation. The main desired properties are defined as being automated, accurate and independent from the availability of training set and operators.

The objectives can be further categorized as following to achieve the main objective and address the problems stated previously:

1. Critically evaluate possible segmentation technique and analyze their applicability in the context of hand bone segmentation.
2. Redefine the eventual purpose of existing histogram equalization and to extend their capability to solve the problem of variability across different radiographs.

3. Propose the utilization of anisotropic diffusion and automate the anisotropic diffusion that able to smooth the irregular and uneven texture within anatomical structures.
4. Instill the human cognitive ability availing of rule-based fuzzy inference system to an adaptive division scheme to search the optimized size for applying the central fundamental algorithm automatically.
5. Design an automated quality assurance scheme to assess the segmentation image and determine the need to restore the lost detail, eliminate unwanted segmented regions or remain as it is.

1.5 Research Scopes

A few scopes associated with database, methodology and result analysis as following are set to define the boundaries of the project and specify the task to be accomplished.

1. Research focuses on designing and performing the proposed segmentation framework on Radius, Ulna and Short Bones (RUS) of gray scale left hand bone radiographs for computer-aided skeletal age scoring system (TW3).
2. Developments of all algorithms and graphs plotting in this book are implemented by using the MATLAB R2012a. The developed image processing algorithm are focused on spatial domain processing instead of frequency domain as the transformed representation is unsuitable in the context of hand bone segmentation where the spatial location of the anatomical structure is of prime importance.
3. A total of 1,400 hand radiographs that are implemented in testing and analyses are acquired from hand bone online database, http://ipilab.org, which comprises of both genders in four populations which are Caucasian, African American, Hispanic, and Asian, ages ranging between 0 and 18, collected from Children's Hospital Los Angeles (CHLA).
4. To have fair comparisons, the performance evaluations of proposed automated segmentation frameworks are conducted in two methods:
   a. Analytical evaluations method is performed on existing methods that are non-automated where direct implementations without human interventions are not possible or are subjective to justify the performance fairly.
   b. Empirical evaluations method is performed on existing methods that are automated where direct implementations are possible and objective to justify the performance.

1.6 Research Contributions

The contribution of this book involves addressing the problems of hand bone segmentation followed by filling in the technical, theoretical and conceptual gap between the basic segmentation methods and the state-of-the-art deformable models segmentation methods to develop an effective segmentation frameworks consisting of several modules that are fully automated and independent from the completeness of training samples and availability of skillful operators. As will be shown, this proposed segmentation framework produces superior segmentations yet it remains computationally feasible.

The main contributions presented in this book are summarized as follow:

1. Extend the comprehensiveness of existing histogram equalization technique by first assessing the current theoretical and technical architecture of existing histogram equalization methods and then contribute the new insight to revolutionize the conventional perception towards the ultimate goal of histogram equalization by proposing the new histogram equalization framework; then, based on the revolutionized insight, a holistic histogram equalization in terms of luminance preservation, contrast, and detail preservation based on the Beta function is developed to preprocess the hand bone radiograph serving the purposes of standardizing and equalizing the non-standardized illumination among radiographs that contain high variations in luminance, improving luminance difference across edge borders in radiographs, reducing variations in luminance difference across edge borders among radiographs and most importantly, enhancing the visual perceptual effect of ossification sites to improve the performance in ossification localization and bone age assessment.

2. Extend the body of knowledge of anisotropic diffusion by exploiting the potential of being fully automatic and adaptive to input radiograph instead of being subjectively tuned by operators to solve the problem of non-uniformity and mitigate the undesired effect of overlapping intensity range. Both contributions below have profound implication for advancing the field of anisotropic diffusion and provide adequate ground for framework that requires autonomous anisotropic diffusion.

   a. Address the problem of manual diffusion strength by designing an automated diffusion strength scheme based on the diffusion coefficient function of speckle reducing anisotropic diffusion (SRAD) grounding on the well-founded statistical theory of the relation between sample variance and global variance. The main strength is its computationally attractiveness and practical applicability.

   b. Address the problem of manual scale selection by designing an automated scale selection. The main strength of which compared to limited existing automated scale selection schemes is that it requires no excessive filtered image before making decision to halt the diffusion iteration.

3. Transform the manual and rigid adaptive division scheme into an automated adaptive quadruple division scheme that embodies human cognitive ability. This transformation is significant not only in a narrow sense of hand bone segmentation, but most importantly, it is of a generic breakthrough in the field of image segmentation. This implicit modeling of human intuition and prior knowledge solve the problem of high dependency on explicit human resources in operating the algorithm. Furthermore, the scheme itself is a building block or

framework for other segmentation algorithm to determine the optimum region size for algorithm implementation.
4. Incorporate quality assurance module in the segmentation framework to evaluate the appropriateness to serve as input for subsequent stages of computerized bone age assessment. The step is important to eliminate over-segmented regions of hand bone and restore under-segmented regions to further improve the quality of segmented hand bone. This concept provides an insight that a procedure for imperfect segmentation framework system that is capable of analyzing the current output and patch up the incompleteness accordingly is needed.

In conclusion to contributions, this book provides the contrary perception to conventional concept that prone to complicating the segmentation algorithm to seek for enhancement in segmentation performance. Instead, the proposed segmentation framework pioneers the insight by postulating that combinations of several customized modules are capable of achieving result that tantamount to result achieved by complicated algorithm or algorithm that demands scarce and limited resources. The conception lies in the strategy to identify, target and analyze the principal adversities that impede the performance of existing methods; then, based on the analyzed result, the contribution is claimed to be advancing the concept of adapting the input information to the central segmentation algorithm (in this book, it refers to unsupervised clustering). This concept is of contrast to conventional segmentation framework that tends to complicate the fundamental segmentation algorithm to adapt the input information. The strength of each module and the idea of concatenating each of them by utilizing the by-product of each module are breaking new ground in the field of automated image processing. These instructive insights embrace the potential to spark attentions and generate new research grounds that will in turn contribute in other fields of applications.

1.7   Book Organization

The organization of the book is summarized as follows.

This chapter introduces the background of research topic, presents the motivation of the research, states the research problems, declares the objectives of the research, specifies the scope and identifies the significance of the research and contributions of the research.

Chapter 2 presents the literature review by providing the critical appraisal of various existing segmentation techniques which include the discussion on the basic concept, mathematical formulation and their strengths and weaknesses in the context of hand bone segmentation for computer-aided skeletal age scoring system. This chapter concludes that the existing segmentation techniques are still technically impractical for fully automated hand bone segmentation.

Chapter 3 presents the design and implementation of the proposed segmentation framework that consist of several modules. Firstly, we describe the identification of the desired properties of segmentation framework as a result of critical appraisal on reviewed literatures to provide direction for the top-down strategy in designing the segmentation framework. Secondly, we describe the proposed extension of current histogram equalization into holistic histogram equalization. Thirdly, we present the extension of anisotropic diffusion and the incorporation of which into the framework. Then, we present the implementation of texture based clustering algorithm inside the module of the proposed adaptive automated fuzzy quadruple division scheme. At last, we present the last module of quality assurance procedure to complete the segmentation framework.

Chapter 4 discusses qualitative and quantitative analysis to justify the motivation and also evaluate the performance of each proposed module and the overall automated segmentation framework. Each experimental result or result analysis is followed by result discussion to interpret the implication of the result to the research topic.

Finally, Chapter 5 summaries the research findings as well as provides suggestions for exploration and extension of future research that might be relevant and important for further development and improvement of the proposed segmentation framework.

References

1. Tanner JM (2001) Assessment of skeletal maturity and prediction of adult height (TW3 method), W.B. Saunders, London
2. Aicardi G, Vignolo M, Milani S, Naselli A, Magliano P, Garzia P (2000) Assessment of skeletal maturity of the hand-wrist and knee: a comparison among methods. Am J Hum Biol 12:610–615
3. Cao F, Huang HK, Pietka E, Gilsanz V (2000) Digital hand atlas and web-based bone age assessment: system design and implementation. Comput Med Imaging Graph 24:297–307
4. Martin DD, Heckmann C, Jenni OG, Ranke MB, Binder G, Thodberg HH (2011) Metacarpal thickness, width, length and medullary diameter in children-reference curves from the First Zürich longitudinal study. Osteoporos Int 22:1525–1536
5. Roche AF, French NY (1970) Differences in skeletal maturity levels between the knee and hand. Am J Roentgenol 109:307–312
6. Chemaitilly W, Sklar CA (2010) Endocrine complications in long-term survivors of childhood cancers. Endocr Relat Cancer 17:R141–R159
7. Heinrich UE (1986) Significance of radiologic skeletal age determination in clinical practice. Die Bedeutung der radiologischen Skelettalterbestimmung für die Klinik 26:212–215
8. Peloschek P, Nemec S, Widhalm P, Donner R, Birngruber E, Thodberg HH et al (2009) Computational radiology in skeletal radiography. Eur J Radiol 72:252–257
9. Tristán-Vega A, Arribas JI (2008) A radius and ulna TW3 bone age assessment system. IEEE Trans Biomed Eng 55:1463–1476
10. Greulich W, Pyle S (1959) Radiographic atlas of skeletal development of hand wrist
11. Tanner J, Whitehouse R (1975) Assessment of skeletal maturity and prediction of adult height (TW2 method)
12. Aja-Fernández S, De Luis-García R, Martín-Fernández MÁ, Alberola-López C (2004) A computational TW3 classifier for skeletal maturity assessment. A computing with words approach. J Biomed Inform 37:99–107

13. Acheson RM, Fowler G, Fry EI, Janes M, Koski K, Urbano P et al (1963) Studies in the reliability of assessing skeletal maturity from x-rays. 3. Greulich-Pyle Atlas and Tanner-Whitehouse method contrasted. Hum Biol; Int Rec Res 35:317–349
14. Martin DD, Deusch D, Schweizer R, Binder G, Thodberg HH, Ranke MB (2009) Clinical application of automated Greulich-Pyle bone age determination in children with short stature. Pediatr Radiol 39:598–607
15. Ontell FK, Ivanovic M, Ablin DS, Barlow TW (1996) Bone age in children of diverse ethnicity. Am J Roentgenol 167:1395–1398
16. Tanner JM, Gibbons RD (1994) Automatic bone age measurement using computerized image analysis. J Pediatr Endocrinol 7:141–145
17. Hsieh CW, Jong TL, Chou YH, Tiu CM (2007) Computerized geometric features of carpal bone for bone age estimation. Chin Med J 120:767–770
18. Pietka E, Gertych A, Pospiech S, Cao F, Huang HK, Gilsanz V (2001) Computer-assisted bone age assessment: Image preprocessing and epiphyseal/metaphyseal ROI extraction. IEEE Trans Med Imaging 20:715–729
19. Thodberg HH, Jenni OG, Caflisch J, Ranke MB, Martin DD (2009) Prediction of adult height based on automated determination of bone age. J Clin Endocrinol Metab 94:4868–4874
20. Jonsson K (2002) Fundamentals of hand and wrist imaging. Acta Radiol 43:236

Chapter 2
Literature Reviews

Abstract In this chapter, we review the techniques of histogram equation, hand bone segmentation and image segmentation techniques. The basic concepts of those techniques will be analyzed and compared. The objective of this chapter is to show to readers that current techniques are not sufficient to address the hand bone segmentation problems due to various restrictions. Hence, this motivates the need for new technique that is capable of segmenting the hand bone so that the segmented hand bone can be established and implemented in the fully automated computer-aided skeletal age scoring system. The proposed novel framework of hand bone segmentation will then be presented in next Chap. 3.

2.1 Introduction

This section provides an overview for traditional or conventional histogram equalization method and segmentation techniques that are usually adopted in applications related to images segmentation to give the reader an overview of the development of these techniques. The fundamental concept of each technique is presented and the pros and cons of each technique are discussed. Besides, the unsuitability of traditional techniques in hand bone segmentation are illustrated in the context of computer-aided skeletal age scoring system by analyzing the nature of the technique and by implementing the analyzed technique in hand bone segmentation. This evaluation and implementation of previous techniques in hand bone segmentation are crucial to motivate the objective and justify the contribution of this book. This section ends with the conclusion that a more advanced technique of hand bone segmentation should be derived instead of using the traditional segmentation techniques.

Y. C. Hum, *Segmentation of Hand Bone for Bone Age Assessment*,
SpringerBriefs in Applied Sciences and Technology, DOI: 10.1007/978-981-4451-66-6_2,
© The Author(s) 2013

2.2 Existing Hand Bone Segmentation

All the computer-aided BAA system require a segmentation stage with the purpose of eliminating the background, noise, soft-tissue region that contains no pertinence of information in the deduction of the skeletal maturity measurement [1–6]. Instead, this undesired information will affect the subsequent stages of computer-aided skeletal age scoring system and in turn deteriorate the result accuracy. However, most of the conventional methods used in this pre-processing stage are either not practical in terms of resources consumption in most of the context or are not yet developed into fully automated framework. Besides, most of the researchers perform the segmentation after obtaining the region of interest (ROI) to reduce the difficulty of segmentation [7, 8]. In fact, the segmentation accuracy and ROI searching ability can be enhanced by employing the algorithm after the hand bone partition from the soft-tissue region. Therefore, As the main pre-processing stages of the computer-aided system, the output accuracy and practicality of segmentation are critical because the eventual outcome of the computer-aided skeletal age scoring system depends on this partition procedure [9].

Substantial studies have been conducted to solve the hand skeletal bone partition problem to exclude it from soft-tissue region and background. Most of the studies involve the employment of threshold that is considered impractical in the hand bone segmentation as the soft-tissue region possesses pixel that have similar intensity to pixels found in spongy bone [1, 3, 10, 11]. Moreover, majority of the studies, implements the active contour model after obtaining the region-of-interest (ROI), [12]. This leads to drawbacks such as the contour sensitive to intensity gradient, depends heavily on initiate position and exhibits inability in extending into concavity [13–15]. Besides, some studies have adopted the statistical analysis to acquire the membership of each pixel i.e., to determine the labeling of the pixels either to the bone or the soft-tissue region [16, 17]. Also, some researchers incorporate various segmentation techniques into the hand skeletal bone segmentation [5, 18]. The summary of the studies are described in the next few paragraphs.

As early attempt, Michael and Nelson [19] proposed a computer–aided diagnosis (CAD) system for BAA that includes pre-processing, segmentation and measurement. The image pre-processing is done via the histogram equalization and then followed by binary conversion of the radiograph and implementing the thresholding using intensity of pixels to eliminate the background by the model parameters. By using the model parameter, the main drawback is the overlapping problem of pixel intensity in bone and background. Moreover, high sensitivity to illumination uniformity and the presence of soft-tissue region locating around the hand bone have further deteriorated the result. Manos et al. [20, 21] proposed a framework for the automatic hand-wrist segmentation; they have implemented a region growing and region merging technique after performing the edge detection during the pre-processing. Amidst this technique, thresholds involve to determine the efficiency of the edge and growing and merging algorithms. Furthermore, region growing result depends heavily on the performance of edge detection. Lastly, the region merging depends on grey level similarity size and connectivity which bear a risk of combining the epiphysis sites that are situated around the metaphysis.

Sharif et al. [22] have conducted a study on bone outlines detection to perform bone segmentation using edge detection depending on the intensity from the Derivative of Gaussian (Drog) before implementing the thresholding technique. The technique of pre-processing proposed by Mahmoodi et al. [13] involve converting the image into binary image and adopting the thresholding technique determined by image histogram to acquire the ROI. The subsequent segmentation of epiphysis residing in the ROI is performed via the active shape model framework. As mentioned, the drawbacks of this method are associated with the sensitivity in uneven illumination and the presence of soft-tissue region. The pre-processing method used in Mahmoodi et al. [23] for segmentation of bone using deformable models and a hierarchical bone localization scheme. The method background removing process is performed only after obtaining the ROI. Mahmoodi et al. [14] adopt binary thresholding to acquire the delineation of the hand, followed by location searching of concave-convex; finally the segmentation is performed by the method of active shape models.

Sebastian et al. [15] segmented the carpal bones from CT images by deformable models, the pre-processing incorporates the strength of various segmentation techniques such as snake models, region-based segmentation, global competition in seeded region growing and also the local competition in region competition. The drawback of this method is that it is complicated and involves intensive computing consumption during the computation of the partial differential equation. Besides, active contour model has been invariably adopted in partitioning the hand bones, the methods c-means clustering algorithm, Gibbs random fields and estimation of the intensity function have been also proposed by Pietka et al. [24]. Also, They suggested to segment the hand bone using the histogram analysis during pre-processing stage [25] by acquiring the histogram peak of pixels intensity followed by identifying the background and soft-tissue region.

Hsieh et al. [26] incorporate adaptive segmentation method with Gibbs random field at the pre-processing stage. Zhang et al. [27] suggest segmenting the carpal by non-liner filter as pre-processing follow by adaptive image threshold setting, binary image labelling and small object removal. However, it involves user-specified threshold and Canny edge detection which are not robust in segmentation. Similarly, Somkantha et al. [5] segment the carpals bone using a combination of vector image model and Canny edge detector. Han et al. [7] propose to implement watershed transform and gradient vector flow (GVF) to perform the segmentation where the performance of watershed transform and GVF depends heavily on edge gradient strength. Tran Thi My Hue et al. [28] propose to implement watershed transform with multistage merging for the segmentation task.

The utilization of the state-of-the-art technique of deformable model such as Active Shape Model (ASM) and Active Appearance Model (AAM) of hand bone segmentation has gained considerable attentions in recent years [29–31]. The strength of this method is that it is well-founded on statistical learning theory. However, the main drawback of this technique is that it is not yet developed into a fully automated system. The initialization of the technique is to delineate the hand bone shape and this thus far is accomplished manually.

2.3 Thresholding

Thresholding is one of the earliest image segmentation techniques, and yet it remains to be the most widely applied segmentation technique attributable to its simplicity and intuitiveness [32]. Thresholding segmentation is normally conducted in spatial domain based on the postulation that both object and background are represented by different range of pixel intensity [33]. Basically, there are three categories of thresholding: Global thresholding, local thresholding and dynamic thresholding [34].

2.3.1 Global Thresholding

Undoubtedly, the simplest method in thresholding techniques to segment an image is through Single Global thresholding: this technique based on the concept that if object in the image and other object or background are mutually exclusive in terms of intensity range, then it could be separated in different partition using a single or multiple values of pixels intensity [35]. In the case of single threshold, it can be represented as follows:

$$f(x,y) = \begin{cases} g_1 & \text{if } f(x,y) < T \\ g_2 & \text{if } f(x,y) \geq T \end{cases} \tag{2.1}$$

where g_i denotes the group of pixels that represents an object or background; if a pixel value is less than T, which is the threshold value, then it is grouped into g_1; if a pixel value is greater than T, then it is grouped into g_2. The $f(x,y)$ denotes the image pixel intensity in 2D gray-scale image in coordination (x,y). The concern of the technique is to classify an image into object and background; this type of grouping is called binarization.

The single thresoholding depends on the T. This T value determines the intensity range of an object and the intensity range of the image background. For instance (if the object is brighter than the background), if an pixel intensity value is greater than the threshold value, then the pixel will be classified as object; for the pixels which possesses intensity value less than or equal the threshold value, they will be considered as background. This kind of thresholding method is considered as 'threshold above'; another type is 'threshold inside' where the object value is in between two threshold values; similarly, another variants is 'threshold outside' where the values in between the two threshold values are classified as background [36].

The accuracy of thresholding technique in segmentation mainly depends on two factors: first factor is the property of the image intensity distribution of both object and background. Thresholding technique performs most efficiently when the intensity of input image has distinct bi-modal distribution without any overlapping range of intensity for object and background [37]. Overlapping range of intensity occurs often due to uneven illumination. Besides, the nature of the object itself can lead to overlapping range in which some regions within the objects in input image has overlapping range of intensity to background.

As mentioned in previous chapter, one of the natures of X-ray hand bone radiograph is its uneven illumination throughout the image; as well as its overlapping range of intensity distribution among soft-tissue region, trabecular bone and cortical bone due to the nature of hand bone and uneven background illumination as well. The illustrations of the effect of uneven illumination and overlapping range of intensity using Global thresholding in hand bone are demonstrated below. Figure 2.1 shows the input image in the illustration and its corresponding histogram. Figure 2.2 shows the resultant segmented image using different threshold, T.

The reasons for the inferior quality of segmented hand bone as shown in Fig. 2.2 can be summarized as follow:

1. Assumption that the whole targeted object (which is the hand bone in this case without soft-tissue region) contains similar intensity range. This is always not true for hand bone radiograph as within the hand bone, there are regions of trabecular bone and cortical bone which have different bone density and hence are represented by different range of pixel intensity values in digital image as shown in Fig. 2.1a.
2. Assumption that the histogram of targeted object and background (dark regions and soft-tissue regions) is perfectly separated into two groups of intensity distributions. This is always not true for hand bone as shown in Fig. 2.1b that the histogram of hand bone radiograph is not bi-modal distributed. This can be explained from the nature of hand bones that are formed by three classes of regions: bone, soft-tissue regions and background instead of two.
3. Assumption that there is no overlapping of intensity range between background and targeted object. This is always not true for hand bone as the some of the

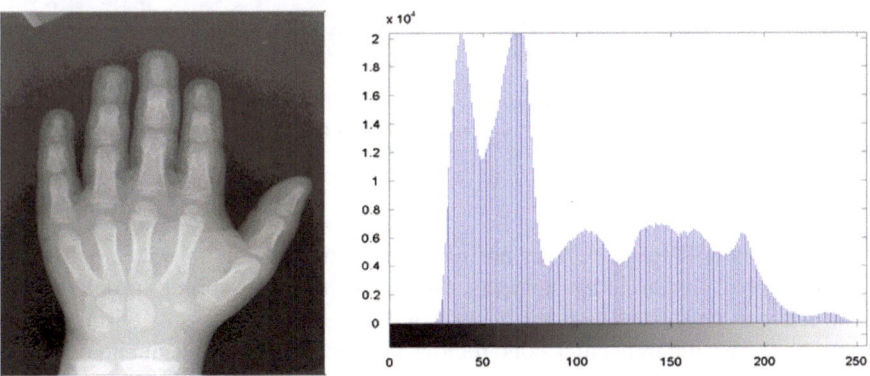

Fig. 2.1 The original hand bone radiograph and its corresponding histogram

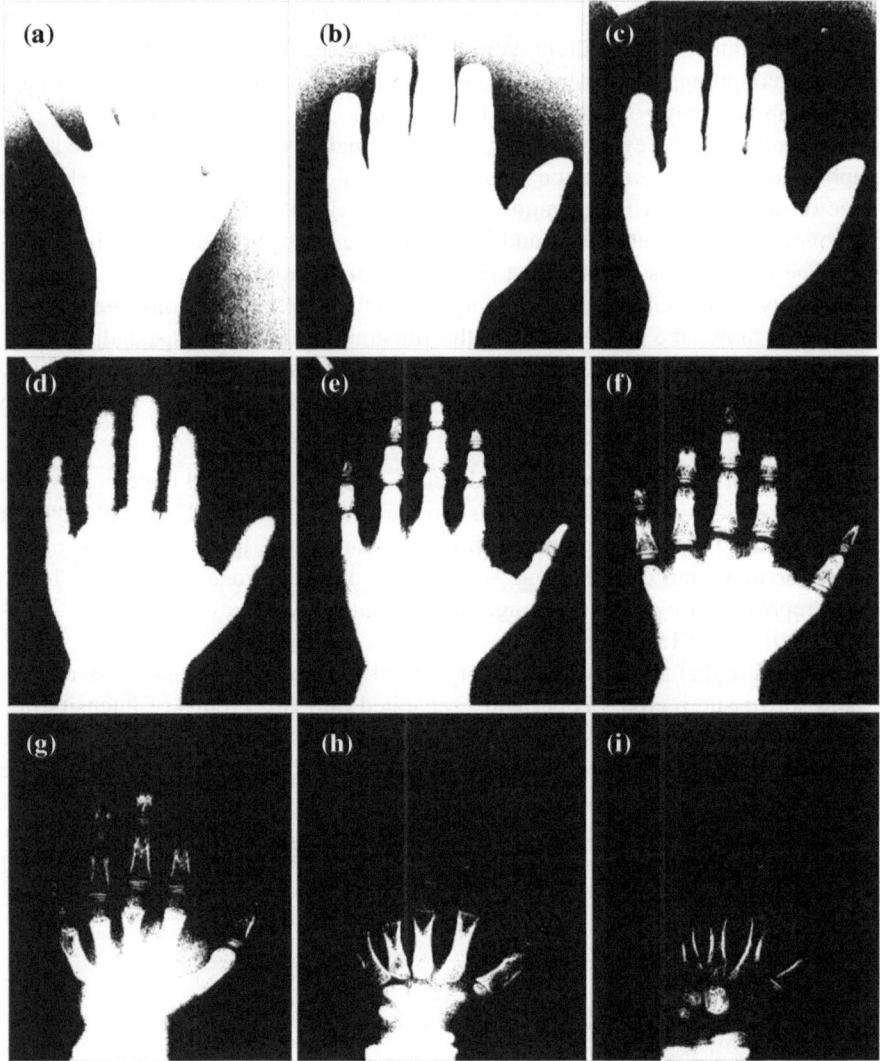

Fig. 2.2 The resultant segmented image in Fig. 2.1 using thresholding with different threshold values T. **a** T = 40. **b** T = 60. **c** T = 80. **d** T = 100. **e** T = 120. **f** T = 140. **g** T = 160. **h** T = 180. **i** T = 200

intensity in soft-tissue regions are identical to the regions in trabecular bones. The Global thresholding neglects this intensitiy overlapping problem. The effects can be best illustrated using Fig. 2.2g–i where the soft-tissue regions together with the trabecular bone regions are simultaneously segmented.

4. Assumption that the illumination is even in input image. This is always not true in for hand bone radiograph as shown in Fig. 2.1a that lower region of hand

bone radiograph has more intense illumination relative to upper region of the radiograph. The Global thresholding neglects this uneven illumination and this affects the segmentation result. The effects can be perceived obviously especially via Fig. 2.2a–c.

Another critical problem of single Global thresholding is the choice of the threshold value to obtain favorable segmentation result [38]. In fact, even the 'best' threshold value is selected, the resultant segmented image in the context of hand bone radiograph and in other medical image processing remain inferior. This fact is inevitable due to the nature of Global thresholding and the nature of hand bone segmentation: only one threshold is selected for thresholding. One improvement for this limitation is by adopting multiple Global thresholding [39]. Multi-level thresholding classifies the image into multiple classes (>2) [40]. The multiple thresholding can be represented as follows:

$$f(x,y) = \begin{cases} g_1 & \text{if } f(x,y) > T_1 \\ g_2 & \text{if } T_1 < f(x,y) \leq T_2 \\ \vdots & \vdots \\ g_{n-1} & \text{if } T_{n-2} \leq f(x,y) < T_{n-1} \\ g_n & \text{if } f(x,y) \geq T_{n-1} \end{cases} \tag{2.2}$$

where g_i denotes group of pixel that represents an object or a background. T_i denotes the threshold value. The $f(x,y)$ denotes the image pixel intensity in 2D gray-scale image in coordination (x,y).

Multiple thresholding might solve the problem arises from the assumption that the input image is of bi-modal type but solve not the problem arises from assumption that the input image is of even illumination. In next sub-section, we would review and examine the local thresholding that is claimed to be more effective in tackling the problem of uneven illumination [41].

2.3.2 Local Thresholding

Local thresholding is segmentation using different thresholds in different sub-images of input image [42]. The input image is firstly divided into a number of sub-images, and then in each sub-image, suitable threshold is chosen to perform the segmentation, this process repeats until all sub-images undergo the Thresholding segmentation. Adopting different threshold in different region of the input image is proven to be more effective than Global thresholding in that it is easier to obtain well-separated bi-modal or multiple-modal distributions in the sub-images and hence it improves the segmentation result [43]. In addition, sub-images are more likely to have uniform illumination. This implies that local thresholding could resolve the problem that arises from the non-uniform illumination [44].

Undoubtedly, local thresholding performs better than Global thresholding in tackling the problem of uneven illumination. However, there are difficulties in applying the technique effectively in hand bone segmentation due to the problems as follow:

1. The problem arises from making the assumption that no intensity overlapping between targeted object and background.
2. The size of each sub-image is difficult to determine. If the size is smaller or larger than it should be, then the result might be even more inferior to using Global thresholding.
3. The size of the sub-images is globally set and is fixed throughout the entire image. Some regions need smaller sub-image whereas some regions need larger sub-image in local thresholding to optimize the segmentation and the computational efficiency.
4. The number of thresholds needed in each sub-image is difficult to determine.
5. The computational cost is higher in comparison with Global thresholding.

The threshold values are difficult to be set manually as the number of sub-images increases [10]. In Global thresholding as well, the threshold value need to be correctly set in order to optimize the result. Single global threshold using is set human inspection. However, comparing with multiple thresholding or local thresholding, automated thresholding is more suitable to decrease repetitive threshold setting by human which is subjective and yet time-consuming.

2.3.3 Dynamic Thresholding

In Global thresholding, each pixel is compared with the global threshold; in local thresholding, each pixel in sub-image is compared with each local threshold which is computed from each sub-image; in dynamic thresholding, each pixel is compared with each dynamic threshold which is computed from sliding a kernel over the input image [43]. One of the popular of dynamic thresholding methods is Nilback method [45].

Niblack method: A w x w kernel moves in the input image from pixel to pixel. Both mean, $m(x, y)$ and standard deviation, $s(x, y)$ in each position of kernel are defined as follow:

$$m(x, y) = \frac{1}{w^2} \sum_{i=x-\frac{w}{2}}^{i=x+\frac{w}{2}} \sum_{j=y-\frac{w}{2}}^{j=y+\frac{w}{2}} f(x, y) \tag{2.3}$$

$$s(x, y) = \sqrt{\frac{1}{w^2} \sum_{i=x-\frac{w}{2}}^{i=x+\frac{w}{2}} \sum_{j=y-\frac{w}{2}}^{j=y+\frac{w}{2}} (f(x, y) - m(x, y))^2} \tag{2.4}$$

The dynamic threshold, $T(x, y)$ is computed via Weighted-Sum operation of the mean and standard deviation and is denoted as follows, where c is a constant:

$$T(x, y) = m(x, y) + c * s(x, y) \tag{2.5}$$

After T is computed, the image undergoes thresholding operation in each pixel. In the case of binarization in 8 bits input image, the segmentation can be denoted as follows:

$$f(x, y) = \begin{cases} 0 & \text{if } f(x, y) < T(x, y) \\ 255 & \text{if } f(x, y) \geq T(x, y) \end{cases} \tag{2.6}$$

Generally, dynamic thresholding performs better than Global thresholding and local thresholding. However, it has similar drawback to local thresholding such that the kernel size is determined manually; the constant in Eq. (2.5) has to be selected manually depending on application. Only suitable selection of kernel size and constant can produce optimum result of segmentation. In addition, dynamic thresholding consumes much more computational resources relative to local thresholding and Global thresholding due to its pixel-wise nature. Besides, in performing the neighborhood operations for dynamic thresholding, the padding problem arises when the kernel approaches the image borders where one or more rows or columns of the kernel are placed out of the input image coordinates.

In next sub-section, the automated threshold value setting techniques which can be applied in both Global thresholding and local thresholding are explored and studied. The implementations of multiple thresholding and local thresholding in hand bone segmentation is illustrated in next sub-section using automated threshold values selection to demonstrate that the sole implementation of these techniques fail to provide good segmented hand bone.

2.3.4 Automated Thresholding

The main technical issue being frequently discussed is the threshold value selection: the decision to determine the threshold value in which the object and the background could be separated as accurate as possible or the decision to select the threshold value so that the object and the background misclassification rate are lowest. The result of Thresholding segmentation process depends heavily on this value. An inaccurate or inappropriate setting of this value will produce disastrous result in Thresholding segmentation.

For the choice of threshold value, basically, there are two main methods: the manual threshold selection and the automated threshold selection. Manually determined threshold value heavily relies on human visual system. Threshold value is selected using visual perception to partition the object from the background; the main drawback of this threshold selection is that it involves human subjective perception towards image quality. Besides, the process itself is extremely time-consuming if the operation involves multiple thresholds. Therefore, it is not practical to determine the threshold value of a large number of images. In short, the manually determined value is not effective.

For automated thresholding method, various methods exist: the simplest method is to utilize the image statistics such as mean, median (second quartile) and first quartile, third quartile, to act as threshold value [46]: this method performs only relatively well in an image that free of noises; the reason is that the noises in the image have influenced the statistic of the image. Typically, if the mean of an image used as threshold value, then it can separate a typical image with object brighter than background into two components; however, while noises exist, the noises have altered the nature that the pixels with intensity greater than mean are belonged to the object. Besides, this kind of thresholding method assumes that the object and the background are themselves homogenous. In other words, the object is a group of pixels containing similar pixel intensity; the background is a group of pixels with similar intensity. This assumption has serious limitation especially in medical image segmentation where the targeted objects such as organs and bone are not inherently homogenous. Besides using simple afore-mentioned statistic in input image, there are other methods to choose the threshold value. In next paragraph, different type of automated thresholding techniques are explored and studied. The Global thresholding and dynamic thresholding using automated threshold selection on hand bone radiographs are shown in Figs. 2.3 and 2.4.

Fig. 2.3 Segmentation result using Global threshold value/values of (**a**) mean (**b**) median (**c**) between first quartile and third quartile (**d**) below first quartile and above third quartile

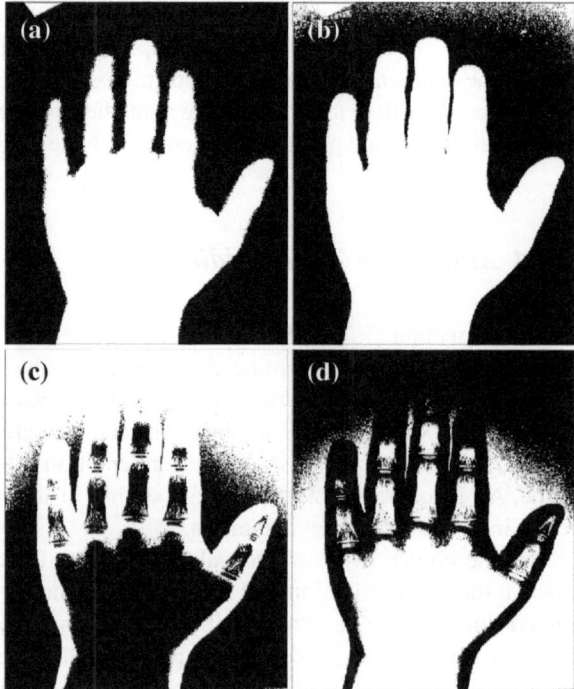

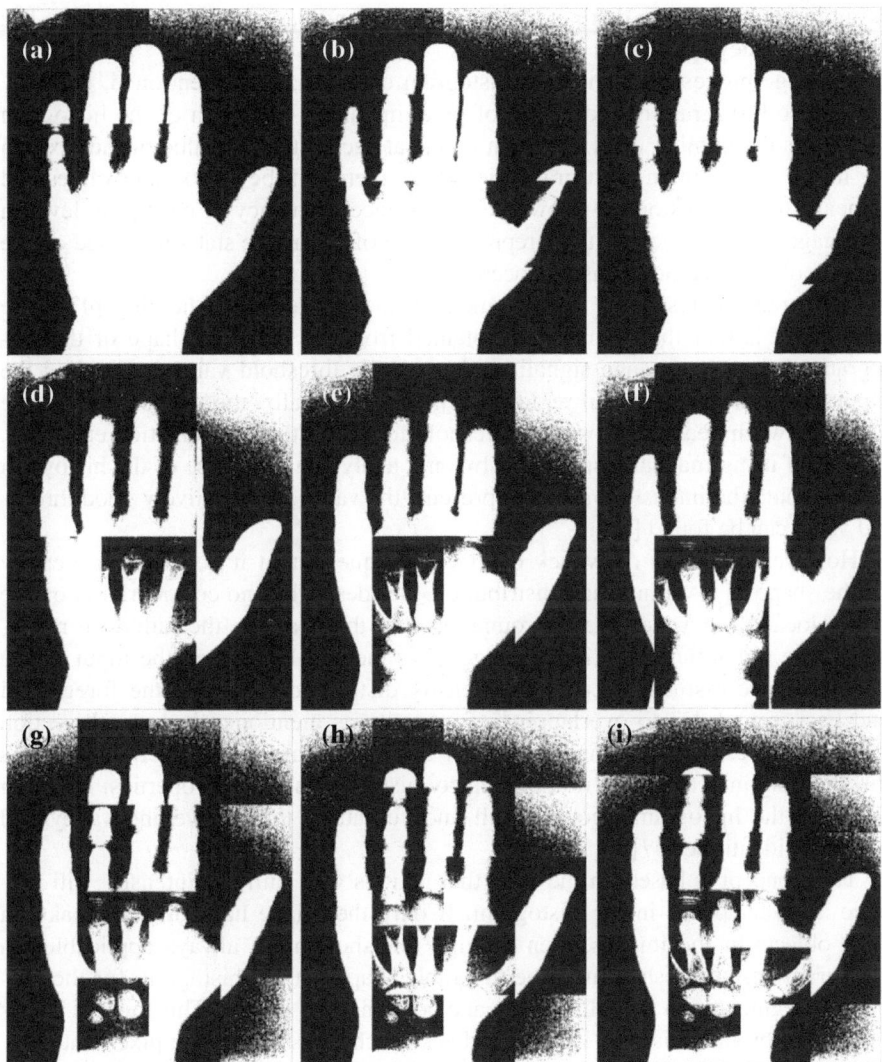

Fig. 2.4 Segmentation result using local thresholding using mean value with different number of division in column and row. **a** 2 columns 3 rows. **b** 2 columns 4 rows. **c** 2 columns 5 rows. **d** 3 columns 2 rows. **e** 4 columns 2 rows. **f** 5 columns 2 rows. **g** 4 columns 3 rows. **h** 4 columns 4 rows. **i** 4 columns 5 rows

Attributable to the limitations of using simple statistics, various more sophisticated type of thresholding method based on different technique in determining the threshold value are proposed: one of the methods is the threshold value selection based on histogram: instead of choosing the mean or median of the image as the threshold value to separate the object and the background, the histogram-based thresholding method determine the threshold value based on the histogram shape assuming that there are distinct range for object and background themselves. The value of a valley point is set as threshold.

In image processing, when the histogram of an image is mentioned, typically it refers to histogram of the values of pixel intensity; the graph of the histogram represents the number of pixels in an image at each intensity value of the pixel in the image. If say in an 8-bit grayscale image, there will be 2^8 possible values and it means that the histogram shows the occurrence frequency of each gray level in the image. In other words, it is a representation of the image statistics based on the number of the specific intensity's occurrence.

Histogram analysis is a popular method in automated thresholding [47]. The postulation is that the information obtained from the physical shape of the histogram of the input image signalizes the suitable threshold value in dividing the input image into meaningful regions [48]. Conventionally, the intensity bin in the valley between peaks is chosen as threshold to reduce the segmentation error rate. Instead of using manual inspections, by only analyzing the shape of the histogram and compute the intensity bin that represents the valley, the relatively good threshold value can be found [49].

However, the main drawback of this technique is that it depends too heavily on the shape of pixel intensity distribution. Besides, it has no consideration on the pixels location and the pixel surroundings and this leads to the failure in recognizing the semantic of the input image. This method fails when the input image does not have distinctly separated intensity distribution between the foreground and background due to overlapping of intensity as mentioned in last sub-section of Global thresholding. This category of automated threshold selection performs thresholding in accordance to the intensity histogram's shape properties. Utilizing basically the histogram's convex hull and curvature, the intervening valley and peaks are identified [47].

This concept is based on the facts that regions with uniform intensity will produce apparent peaks in the histogram. If only the image have distinct peaks on each objects in the images, then multiple thresholding is always applicable via histogram-based thresholding. The favorable shapes of the histogram for the purpose of segmentation are tall, narrow and contain deep valleys. This method is less influenced by the noise but it has drawbacks such as assuming the pixels intensity range of the object and background has a certain degree of distinction. If the image has no distinct valley point in the histogram, this method would fail to separate the object from the background. The main disadvantage of this histogram-based thresholding method is the difficulties they meet when they have to identify the important peaks or valleys in the image used for segmentation and classification.

Selecting the threshold value according to the result of clustering analysis is one of the major automatic thresholding methods. Depending on the type of clustering analysis, various variations of this category of threshold selection are proposed. One of the most widely implemented methods is Otsu method [50, 51]. The author suggested that the optimal threshold is the intensity point that able to divide the input image into two groups of intensity when the intra-class variance is minimum or the inter-class variance is maximum [52].

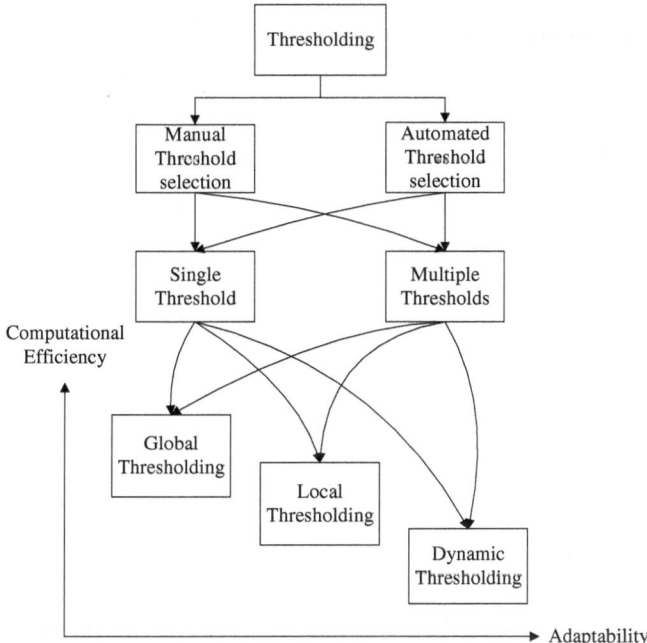

Fig. 2.5 The taxonomy of Thresholding segmentation

Various Thresholding segmentation techniques have been reviewed in this sub-section. Each category of thresholding with different schemes of threshold selection and number of thresholds has different capability in segmentation task. Generally, Global thresholding has the highest computational efficiency followed by local thresholding and dynamic thresholding. Global thresholding, however, has the lowest adaptability to deal with uneven illumination, followed by local thresholding and dynamic thresholding. The experiments on hand bone segmentation using aforementioned thresholding methods showed that thresholding methods can hardly produce promising resultant segmented hand bone; Taking no consideration of object's features in thresholding technique is probably the reason. Therefore, in next sub-section, the edge-based segmentation methods would be explored. Lastly, the taxonomy of Thresholding segmentation is summarized using the Fig. 2.5.

2.4 Edge-Based

Thresholding segmentation discussed in previous sub-section postulates that foreground and background can be distinguished by certain range of pixels intensity. edge-based segmentation, differently, postulates that foreground and background are distinguished by a closed boundary formed by meaningful edges. Edges are

Fig. 2.6 General framework
of edge-based segmentation

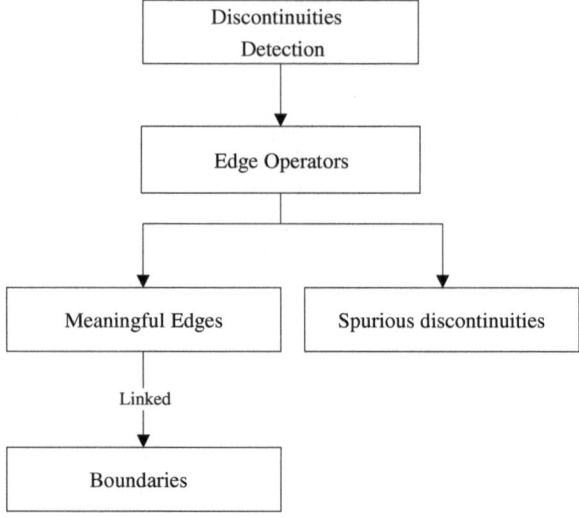

pixels which undergo abrupt changes in gray level intensity, known as discontinuities, defined by derivative values that surpass a pre-selected threshold. However, not all detected discontinuities are meaningful in the context of segmentation [53]. The meaningful discontinuities are termed as 'edges', which are a set of pixels that appertain to the boundary of objects. All these edges, subsequently, are connected and linked to constitute closed boundary; pixels within the closed boundary are then labeled as objects. This type of labeling is referred to edge-based segmentation [33].

Therefore, edge-based segmentation usually deals with two major technical problems [54]: (1) local edge feature; the definition of discontinuities and meaningful edges, such as the features and threshold adopted to define discontinuities. (2) Global edge linking; the linking procedure connects detected binary edge pixels to form linked edges. In this sub-section, some classic edge detectors would be described briefly and several classic edge-based segmentation algorithms are explored (Fig. 2.6).

2.4.1 Edge Detectors

Edge refers to pixels within digitalized image with large gradient, or in layman terms: image points that undergo sharp variations. Edges detection is an operation that converts image into a set of meaningful curves that exhibit certain characteristics or features. This operation is critical in filtering out the meaningless information while retaining the vital objects structural properties.

Edge detectors are kernels used in edge detection operation to measure each pixel's neighborhood and quantify the discrete differentiation of the edge transition include the degree of intensity changes and direction of the changes.

Specifically, the edge detection operation is a convolution operation [55]. Common edge detectors are Robert edge detector, Prewitt edge detector, Sobel edge detector, and Canny edge detector.

Canny edge detector [56] is a multistage algorithm instead of a pair of kernels such as Prewitt and Sobel edge detector [57]. In fact, those primitive edge detectors are implemented in one of the stages in Canny edge detection algorithm. This famous edge detection is proposed by John. F. Canny in [56] to emphasize a set of criteria: (1) True edges should be detected while there is no response to non-edges; (2) The detected edge should be well localized; (3) There is only single response to each edge pixel. To achieve the aforementioned criteria, four main stages of Canny edge detection algorithms are designed by Canny as follow [58, 59]: Noise reduction, gradient computation, non-maximum Suppression and edge tracing via hysteresis thresholding.

2.4.2 Edge Linking

Edge detectors discussed in previous sub-sections produce output image with edge pixels. These edge pixels require a linking process to become a continuous connected curve [33]. Thus edge detector are normally followed by edge linking which is a process of connecting these detected edge pixels into curve that characterizes the object boundaries. The process is important to fill in the gaps due to occlusion, noise or shading in original binary image. Thus, a variety of edge linking approaches is discussed below.

Not all pixels in image can be categorized as edges pixels but only pixels that undergo abrupt intensity changes, sought by aforementioned edge detectors; not all detected edges, nonetheless, are meaningful edges but only pixels which are located at objects boundary. Therefore, to link those meaningful edges pixels into connected boundary, firstly, identify whether a group of pixels are lying on the boundary of specified shape or to detect the presence of subsets of pixels that are almost collinear or collinear [60]. Suppose that subsets of pixels that located on a straight line require to be identified; one possible solution is to search all the lines constructed by every pair of pixels; then label pixels that are adjacent to the specified line as desired subsets of pixels. This operation requires traversing the entire search space implying that it is computationally prohibitive as the size of the search space increases.

With computational feasibility as motivation, Paul Hough [61], in 1962, proposed and patented a technique to transform a straight line in digitalized image space into pictorial representation of a single point in feature space via two parameters of straight line. The basic mechanism of classical Hough transform in straight line recognition is to firstly obtain a binary image using edge detector discussed previously; the following step is to construct the parameter space of the straight line of each edge pixel and subdivide the parameter space into accumulator cells; then, compute the counts for each accumulator cell to detect high concentrated pixel in parameter space [62]; finally, analyze the relationship among detected pixels within each cell to determine whether to link the corresponding line segments in image space into straight line. The central concept is to represent a shape using the shape's parameter as features instead of image pixels in image [63].

Hough transform have been widely applied in edge-based segmentation as linking algorithm attributable to its two special capabilities over other approaches: First, the tolerance towards incompleteness along the boundary line due to occlusion and shape variations; second, the parallel processing capability which is prominent in real-time application [64]. The main idea of Hough transform to determine if a group of pixels are of certain specified shape using shape parameters instead of using pixels connectivity contributes to the capability to certain level of tolerance towards noises [65]. Besides, the mechanisms of Hough transform to combine independent evidence by analyzing the spatial similarity and the intensity of parameter peaks contributes to its ability of recognizing partially deformed shapes. The second capability is due to the nature of Hough transform in treating each pixel in image space independently [66]. This makes the parallel processing by using multiple processing units become possible, implying that real-time application can be realized. Another added advantage of Hough transform is its capability to detect several specified shape simultaneously by accumulating evidence produced by each shape in terms of distinct cluster or peak found in the accumulator array [67].

Despite the discussed advantages, the classical Hough transform is by no means a generalized or an optimized shape detector due to several limitations. The disadvantage of the classical Hough transform is that it is constrained by certain class of analytically defined shape [68]. Besides, the input image for the transform is limited to only binary images after being processed by edge detectors. Also, the resultant connected lines and shape are limited to skeleton lines with one-pixel thickness. Lastly, the principal limitation of Hough transform is that it consumes large space storage and requires high computational resources; the computational cost increases exponentially with the parameters dimensionality. In next paragraph, various improvements have been devised to tackle the aforementioned limitation.

Numerous improvements to Hough transform through generalization of various properties have been proposed over years since its first appearance to address the aforementioned limitations [69–73]. In 1971, Duda [60] simplified the computational complexity of Hough transform by using parameters of radius and angle instead of slope and intercept and further extends Hough transform into more general analytically defined curved shape such as circle. In 1981, The Hough transform was then further generalized using principle of template matching for applications in which the features of the shapes are not possible to be described using simple analytic mathematical equations [74]. Shapiro and Stockman [36] proposed the generalization of Hough transform for multiple level images instead of binary images. Hough transform is then generalized to detect think lines [75].

The traditional edge-based segmentation techniques have been reviewed. Edge-based segmentation assumes that the object in image has distinct boundaries. This assumption is not true in the context of hand bone due to the smooth transition, uneven illumination and low contrast, especially on the distal phalanges. Besides, the edge detectors are sensitive to noise and therefore it is not reliable enough to dispose spurious edges that are not part of the anatomical boundaries of hand bone. In addition, the edge linking process heavily relies on the precursor image supplied from previous stages of edge detector processing. Lastly, the information within the object boundaries is not fully utilized to produce more promising result of segmentation. Therefore, in the next sub-section, segmentation that makes use of region information to group similar pixels with respect to particular attributes into meaningful regions that represent the object are discussed. This region-based interpretation is important as coherent regions that invariably correspond to objects in the image.

2.5 Region-Based

The edge-based segmentations discussed in the previous sub-section attempt to perform object boundaries extraction in accordance to the identified meaningful edge pixels. Region-based segmentations, on the contrary, seek to segment an image by classifying image into two sets of pixels: Interior and exterior, based on the similarity of selected image features. In this sub-section, several classic methods belong to this category are explored and studied.

The region based segmentation is based on the concept that the object to be segmented has common image properties and similarities such as homogenous distribution of pixel intensity, texture and pattern of pixel intensity that is unique enough to distinguish it from other object [33]. The ultimate objective is to partition the image into several regions where each region represents a group of pixels belong to a particular object.

2.5.1 Seeded Region Growing

One of the famous region methods is seeded region growing; this method grows from seeds which can be regions or pixels, then the seeds expand to accept other unallocated pixel as its region member according to some specified membership function [76]. The details of both regions growing method are illustrated in subsequent paragraphs.

Suppose a few groups of pixels or a few pixels are chosen as seed regions or seed pixels to expand to a region coherently, denoted by R_i. Let $M\,(x,y)$ depicts set of all unallocated pixels in the image, described by the mathematical expression as follows [77]:

$$M = \left\{ (x,y) \notin \bigcup_{i=1}^{n} R_i \,\middle|\, N\left[(x,y)\right] \cap \bigcup_{i=1}^{n} R_i \neq \varnothing \right\} \tag{2.7}$$

where n denotes total number of regions, R_i. Where $N[(x,y)$ denotes the immediate neighboring pixels of each element. This indicates that a pixel can only be considered as the neighboring pixel of a region if it fulfils the requirement that its neighboring pixel is overlapping with the region and at the same time it is not yet being allocated.

The expansion depends on a defined similarity measure $S(I,J)$ where I and J are adjacent pixel. This similarity measure is designed such that the score is higher if both I and J have common desired image characteristics and thus they should be grouped together; this process iterates until convergence. For instance, if the common characteristic is intensity value, the membership determination in 2D case should be set as follows where T denotes the threshold:

$$M(x,y) \in R_i \text{ if } S(M(x,y)) \leq T \qquad (2.8)$$

In comparison with deformable model based segmentation, region based segmentation is considered relatively fast in terms of computational speed and resources. Besides, it is certain that segmentation output is a coherent region with connected edges. Simplicity in terms of concept and procedures is an advantage of region growing for immediate implementation [78].

Region based segmentation is insensitive to image semantics; it does not recognize object but only predefined membership function [79]. Besides, the design of the region membership is as difficult as setting a threshold value; region based segmentation is unable to separate multiple disconnected objects simultaneously.

The assumption that the region within a group of object is homogenous has low practical value in hand bone segmentation due to the fact that the bone is formed by cancellous bone and cortical bone that has high variations on texture and intensity range. Besides, in the presence of noise or any unexpected variations, region growing leads to holes or extra-segmented region in the resultant segmented region and thus has low accuracy in certain condition [80]. The number and the location of seeds and membership function in seeded region growing, as well as the merging criteria in split-merge region growing, which will be discussed later, depends on human decisions which is subjective and laborious.

2.5.2 Region Splitting and Merging

Another famous region growing methods is the split and merge algorithm; split and merge is an algorithm splitting the image successively until a specified number of regions remain [81]. To perform the split and merge region growing algorithm, firstly, the entire image is considered within one region. Then the splitting process begins in the region in accordance to the homogeneity criterion; if the criterion is met, then it splits [33]. This splitting process repeats until all regions are homogenous. After the splitting process, the merging process begins. Initially, comparison among neighborhood regions is performed. Then, the region merges to each other according to some criterion such as the pixels' intensity value where regions that are less than the standard deviation are considered homogenous.

The essential concept of region-based segmentation has been reviewed. The purpose is to identify coherent regions defined by pixel similarities. The main challenge of this type of segmentation is often related to the pixel similarities: what are the features that should be adopted as similarities measurement and how the thresholds of chosen features should be set in defining the similarity. The selection of features is difficult as they depend on application. For example, if the targeted object is not a connected object, pixel intensity is not suitable as pixel similarities measurement. The setting of threshold is another tricky challenge as it manipulates the trade-offs in terms of flexibility. For example, if the threshold is set too low, the inferior effect of over-segmentation occurs because pixels easily surpass the threshold leading to larger coherent regions than the actual objects; if the threshold is set too high, otherwise occurs. Region based segmentation unable to segment objects that contain multiple disconnected regions and therefore, in the context of hand bone segmentation, applying only region-based segmentation is inappropriate as children hand bone of different ages for BAA involve different number of bones regions.

2.6 Hybrid-Based

As previously discussed, the Thresholding segmentation, edge-based segmentation and region-based segmentation have their own merits and each in their own way has made an important contribution. Therefore, it is assumed that if all these attributes together are combined, a better segmentation result can be obtained. In the following, watershed algorithm based on embodying the basic concepts of previous discussed thresholding, region growing and edge detection, is discussed. The main concept is to interpret the image topographically using 3D visualization on the image which refers to the two spatial coordinates and pixel intensity.

2.6.1 Watershed Segmentation

Watershed Segmentation is a hybrid-based segmentation that bases on mathematical morphology. An input image in watershed Segmentation is represented topographically as spatially distributed terrain altitudes in accordance to the numerical value of pixel intensity. Such image representation is useful to visualize the notions of watershed as light area, catchment basins and minima as dark areas. The idea of using watershed transformation is first introduced by Digabel and Lantuéjou [82] using binary image. It was then extended to grayscale images based on immersion analogy by Beucher and Lantuejoul [83]. Further study about the practical and algorithmical aspects of watersheds when it combines morphological tools are performed by Vincent and Soille [84] to serve as the basis of modern watershed Segmentation algorithm. The watershed Segmentation in following paragraphs are described briefly [33].

Three group of point need to be defined to understand the algorithm. First group of points refers to points that form the regional minimum. The second group of points, namely the 'catchment basin' refers to points that if a drop of water fall on them, the drop of water will certainly fall to a single minimum. The third group of points, namely the 'watershed lines', refers to points that if a drop of water fall on them, the drop of water will fall with equal likelihood to more than a single minimum. Usually the third group of points forms lines on the topographic surface as ridges to delineate the catchment basin [85].

The segmentation task can be accomplished if the watershed lines can be found. The idea of searching the watershed line is to imagine, given an image, which can be visualized as a topologic surface, and water is added to flood the topologic surface at uniform rate. The water level keeps rising until the water is about to fully cover the catchment basin, then imagines a dam is built to prevent the water from merging the catchment basin. This process repeats until only all the top of the dams of all the catchment basin remain above the water level. These top of dams form boundaries which are watershed lines. These lines are the desired result of watershed Segmentation algorithm.

The classic watershed Segmentation encounters problem of over-segmentation stems from spurious minima. Therefore, various improvement strategies have been published in terms of the algorithm efficiency [86], incorporation of other techniques such as multi-resolution [87, 88], wavelet analysis [89] or both [90], utilization of prior information such as shape and appearance [91], combination with artificial neural network [92]. Despite various improvements, the watershed remains corresponding to image gradients that are always affected by noises within the image that lead to spurious edges. This in turn leads to over-segmentation. Besides the watershed algorithm require threshold setting to determine the object gradients. This thresholding is virtually of little avail in hand segmentation due to the uneven illumination and various gradient definitions for anatomical structure edges. In the context of automated hand bone segmentation, Fig. 2.7 demonstrates that the watershed algorithm is not robust enough to solve the problem of noises, undistinguishable anatomical structures and uneven illumination. The reasons stem from the inherent property of watershed Segmentation as it requires thresholding in converting the input image into binary image or gradient image as pre-processing [93, 94]. Most importantly, the hand bone for bone age assessment encompassing different age group consists of multiple numbers of bones and this objects inherent property simply fails any type of watershed algorithms. The result in Fig. 2.7 illustrates that watershed algorithm is not efficient in segmenting the hand bone.

Watershed Segmentation has been reviewed. This technique resolves some inferior effect of conventional edge-based segmentation: dependency on gradient, resultant thick lines, detected edges pixels are far apart. Unfortunately, this method is too sensitive towards noise and irrelevant changes in pixel intensity leading to over-segmentation. However, this drawback can be greatly improved by using markers found by features associated with background or object in watershed-processed image. Besides, this algorithm usually processes edge

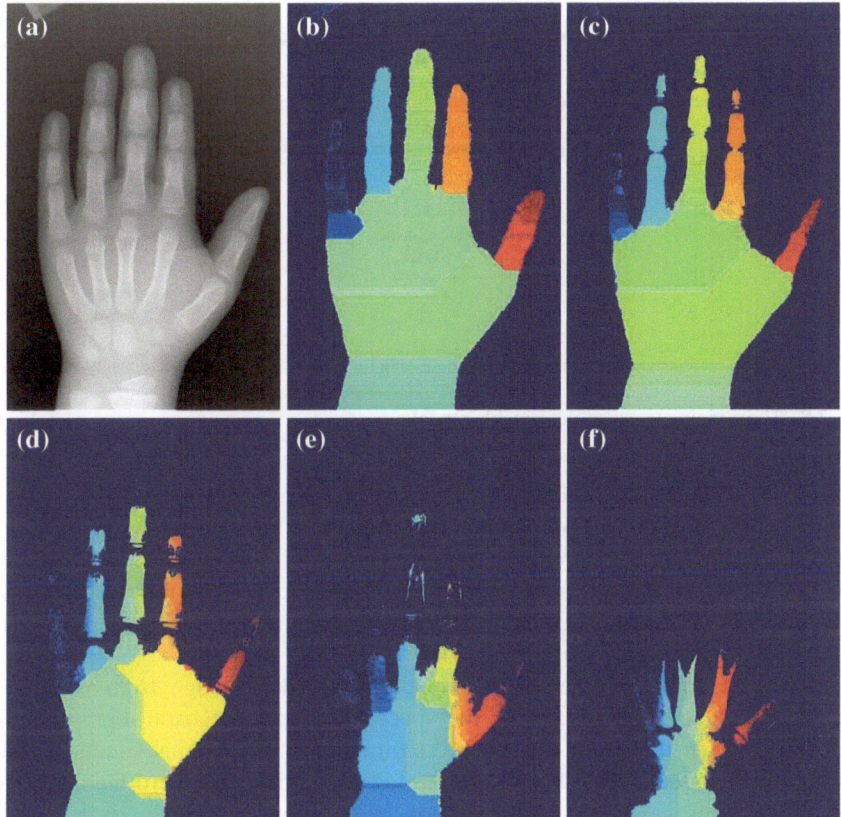

Fig. 2.7 The segmentation performance of watershed algorithm using different value of threshold and result in RGB image showing each labeling group. **a** Original image. **b** threshold = 100. **c** threshold = 120. **d** threshold = 140. **e** threshold = 160. **f** threshold = 180

enhanced image or gradient image; hence, it has the similar drawbacks as mentioned in edge detector sub-section. In general, this hybrid-based segmentation demonstrates that a relatively more robust segmentation technique can be obtained by combining the strengths of segmentation algorithms from different categories.

2.7 Deformable Model

Deformable model refers to classes of methods that implement an estimated model of the targeted object using the model constructed by the prior information such as the texture and shape variability of specific class of object as flexible 2D curves or 3D surfaces. In 2D cases, these curves deform elastically to by satisfying some

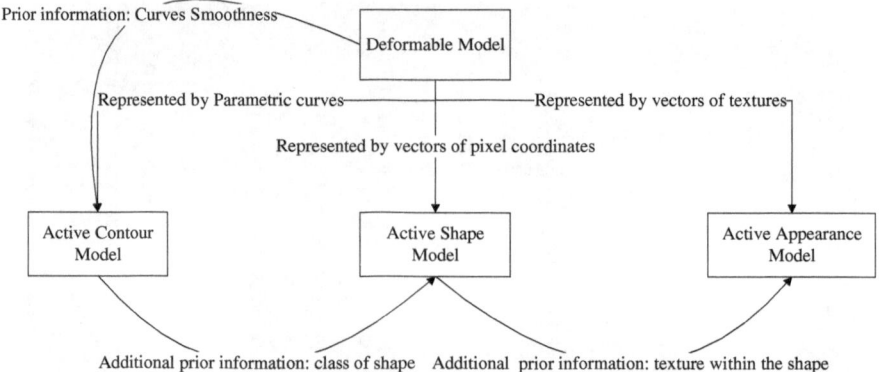

Fig. 2.8   Three main classes of deformable model

constraints to match the borders of the targeted object in a given image. The word
'active' stems primarily from the nature of the curves in adapting themselves to fit
the targeted object. There are three main classes of deformable model: active contour model, active shape model and active appearance model. Each of the classes
differs mainly in the aspect of the incorporated prior information and curves repre-
sentations as illustrated in Fig. 2.8.

Deformable models assemble the mathematical knowledge from physics in
limiting the shape flexibility over the space, geometry in shape representation,
and optimization theory in model-object fitting [95]. These mathematical foun-
dations work together by playing their roles to establish the deformable model.
For instance, the geometric representation with certain degree of freedoms is to
cover broader shape changes; the principle in physics, in accordance to forces
and constraints, controls the changes of shape to permit only meaningful geometric flexibility; optimization theory adjusts the shape to fulfil the objective function constituted by external energy and internal energy; the external energy is
associated with the deformation of model to fit the targeted object due to external
potential energy, whereas, the internal energy constrain the smoothness of the constructed model in terms of internal elasticity forces.

2.7.1 Active Contour Model

Kass et al. [12] proposed active contour model or known as 'snake' as a potential
solution to segmentation problem [96]. From the perspective of geometry, it is an
embedded parametric curve represented as $v(s) = (x(s), y(s))^T$ on image plane
$(x, y) \in R^2$, where x(s) and y(s) denote coordinates functions, and $s \in [0,1]$ denotes
the parametric domain. A snake in this context illustrates an elastic contour that
fits to some preferred features in image. The shape function of the contour that

embodies 'elasticity of contour' and the 'preferred features in image' are determined by the functional as follows:

$$E[v(s)] = E_{int}[v(s)] + E_{ext}[v(s)] \qquad (2.9)$$

This functional of Eq. (2.9) represents the contour energy or the curve spline bending energy. The task is to find the contour function, $v(s)$ that minimizes the functional $E(v(s))$. The first term of Eq. (2.9) represents the internal energy:

$$E_{int}[v(s)] = \int_0^1 w_1(s)\left|\frac{\partial v(s)}{\partial s}\right| + w_2(s)\left|\frac{\partial^2 v(s)}{\partial s^2}\right| ds \qquad (2.10)$$

This internal energy characterizes the deformation of the elastic contour at any point s where $w_1(s)$ denotes the weight function that manipulates contour tension and $w_2(s)$ denotes the weight function that manipulates contour rigidity. These tension and rigidity control the extent to which the contour can deform: Increasing tension tends to reduce the length of the contour and hence the contour can get rid of irrelevant ripples. Increasing rigidity tends to reduce the flexibility and hence the contour can be smoother.

The second term of Eq. (2.9) represents the external energy that relates the contour to the image plane:

$$E_{ext}[v(s)] = \int_0^1 I(v(s)) ds \qquad (2.11)$$

The external energy evaluates the matching between the targeted object's boundaries and the contour. This evaluation is based on certain feature image of interest such as gradient image. If the v(s) located on the edges, then the external energy in Eq. (2.11) at the point s is low. Hence, it is obvious that the minimum external energy occurs when all the points s are located exactly on the detected edges.

In short, the internal energy controls the behavior of curve itself whereas the external energy determines how well the curve matches the features of interest; both of them counterbalance each other. The minimum of the summation of both energies produces a smooth contour that matches the feature of interest, most commonly refers to edges of targeted object and thereby performs the segmentation.

To apply active contour model in segmentation, first, establish the initial location of point s in image planes adjacent to targeted object. These points collect 'evidence' locally in their territories and feedback to the contour energy. Next, search the update of each point using local information by solving the Euler–Lagrange equation when the contour is in equilibrium according to calculus of variation. Conventionally, numerical algorithm is applied to solve the equation in discrete approximation framework. Lastly, these steps repeat until stopping criteria has been achieved.

The advantages of active contour compared with previously discussed methods:

1. Process the image pixels in specific areas only instead of the entire image and thus enhances the computational efficiency.
2. Impose certain controllable prior information.
3. Impose desired properties, for instance, contour continuity and smoothness.
4. Can be easily governed by user by manipulating the external forces and constraints.
5. Respond to image scale accordingly with the assistance of filtering process.

Disadvantages of classical active contour model:

1. Not specific enough to be implemented in specific problem as the shape of the targeted object is often not recognized by the algorithm.
2. Unable to segment multiple objects.
3. High sensitivity to environmental noises in image.
4. High dependency towards intensity gradient along the edges.
5. Do not consider the region information of the targeted objects.
6. High dependency on initial guessed point location. If the initial snake is not sufficiently close to targeted object boundaries, then points in snake can hardly attach the boundaries.
7. Difficult to grow into concavity.
8. Do not have a global shape controller that constraints the shape of contour from deviating from allowable shape of the targeted object.

To sum up, the regularizing terms adopted in active contour model is useful in stabilizing the contour but the robustness is limited as the imposed constraints generally tend to smooth and shorten the contour unless stronger external energy is involved; this scheme is often too general and inadequate. Therefore, a more specifically designed scheme that capable of incorporating more finely tuned prior knowledge about the class of targeted object is required.

2.7.2 Active Shape Model (ASM)

ASM is a model founded on statistical theory where the variations of the shape of the objects can be captured via training procedure using labeled object's contour in the image in set points representation. Activating the trained contour will deform the contour fitting the targeted object in the image. Cootes et al. [97] developed the model. Generally, it works by searching the best position of initial points that are surrounding the object, and then updating these positions until the stopping criteria are achieved through iterations. Ever since the technique is proposed until recently, it has been extensively applied in various fields such as facial recognition [98–100], object tracking [78, 101–104] and medical image processing [11, 105–107].

The first step in ASM is to establish a shape model by a set of shape examples from training images. This process is known as the training phase of ASM. The purpose of this process is to build a typical targeted shape by analyzing various shapes variations over the training images. The process begins with placing landmarks to define the shapes, followed by aligning the shapes to standardized form, and finally, build a statistical model based on the parameter.

'Landmarks' are points that best describe the shape of the object. The process of placing the landmark is a labeling procedure. 'Landmarks' placement can be perceived as feature extraction. These points serve a purpose: as input to the subsequent algorithm to represent the shape of the object. The selections of locations for these points are crucial. The common guidelines while placing the points or 'landmarks' are as follow:

1. Place the points on the objects boundaries.
2. Place the points evenly along the boundaries.
3. Place the points on the same location in every training image.
4. Place the points on corners or high curvature edges of objects boundaries.
5. Place the points on 'T' junctions between boundaries.
6. Place intermediate points between main landmarks.

After placing the 'landmarks', all the 'landmarks' coordinates are collected and vectorized as an n-point column matrix, X, as presented in equation

$$X = (x_1, x_2, \ldots, x_{n-1}, x_n, y_1, y_2, \ldots, y_{n-1}, y_n)^T \qquad (2.12)$$

The users have to specify the number of 'landmarks' and the location of 'landmarks'. These choices are of essence; incorrect decisions are likely lead to disastrous result in latter stages.

The main stages of active shape model are explained briefly as follow:

(a) Model Landmarks Placement

After defining the 'landmarks', they have to be further processed. The processing of these points is known as alignment, a filtering procedure of translational, rotational and scale effect of the shape. This process aligns the landmarks/shape by using training images in order to cope with the non-standard shapes (different orientation, size and position) in training images. The flowchart summarizing alignment procedure is shown in Fig. 2.9. The purpose of aligning the shapes in training images is to assure that all the shapes are invariant in terms of orientation, size and position so that the true shape representation is obtained. These invariants are vital to enable comparison among different normalized shapes. In other words, the shapes in test images require transformation before the training process begins: the size invariant is achieved by scaling operation whereas the operations of rotation and translation deal with orientation invariant and position invariant respectively. This shape invariant is to ensure that all shapes in training images possess approximately same position and same size in x–y plane, as well as to ensure that all shapes are different by minimum distance; only then the shapes in test images are ready to be learnt by the subsequent algorithms.

The Scaling, translation and rotation transformation are known as similarity transformation. The similarity transformation can be expressed mathematically as

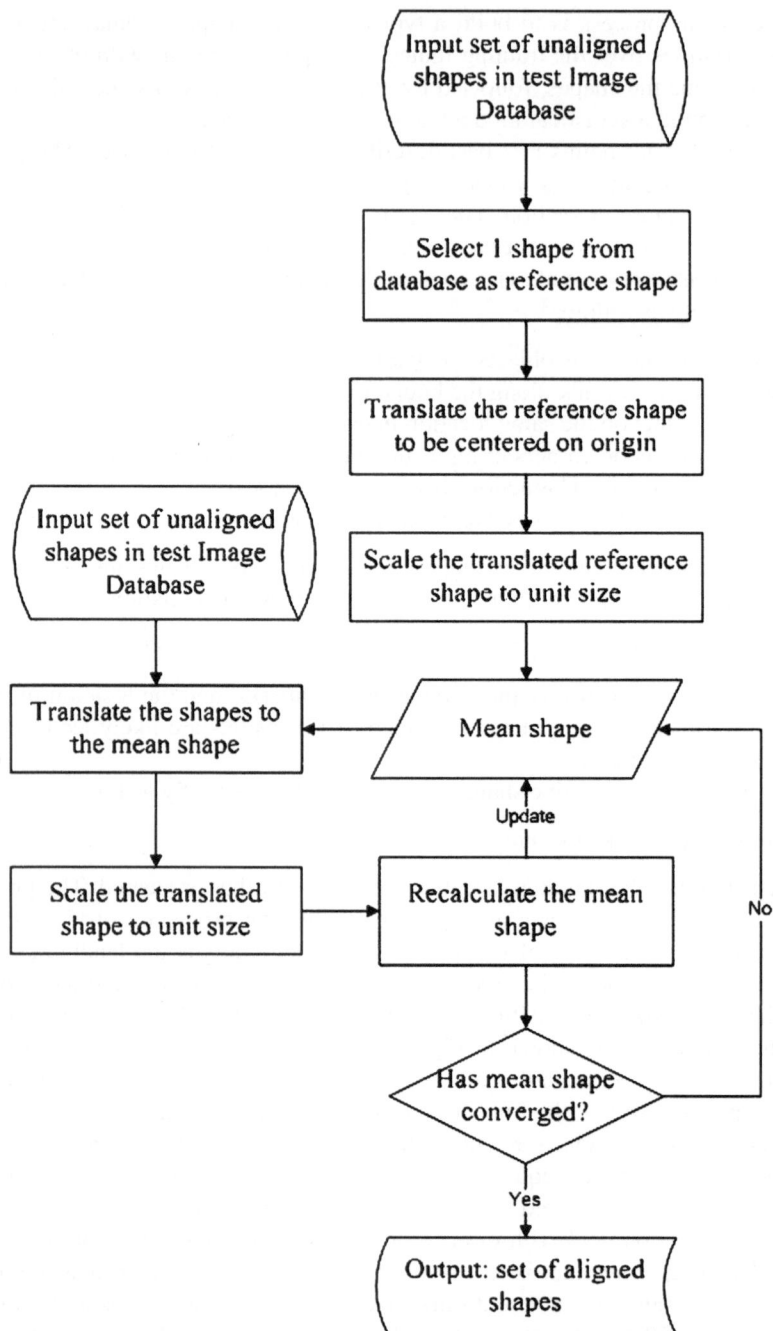

Fig. 2.9 The alignment procedure of ASM

in Eq. (2.13). This equation represents the operations of scaling the point (x, y) by s, rotating the point by (x, y) and translating point (x, y) by (x_t, y_t). The flowchart of the process is shown in Fig. 2.9.

$$T \begin{pmatrix} x \\ y \end{pmatrix} = \begin{pmatrix} x_t \\ y_t \end{pmatrix} + \begin{pmatrix} s\cos(\theta) & -s\sin(\theta) \\ s\sin(\theta) & s\cos(\theta) \end{pmatrix} \begin{pmatrix} x \\ y \end{pmatrix} \tag{2.13}$$

(b) Profile Model

The profile model defines the desired image information around the landmarks region. This information is utilized to model the expected image characteristic around each landmark; the expected characteristic is modeled through various training examples during the training phase; the most commonly adopted characteristic is the edge strength of the targeted image object's boundary; the landmark examines the edge strength along normal to the object boundary edge and moves toward the strongest object boundary edge. Generally, during the matching phase, the landmark adjusts its current position and makes displacement around its surrounding region to search for the optimized position accordingly to the profile model (these adjustments are constrained by shape model which will be discussed later). After every landmark moves to its temporary position accordingly, the overall shape formed by connecting these landmarks is known as the suggested shape. This process of searching and adjusting to form temporary shape iterates until convergence to obtain the final shape.

(c) Shape Model

Shape model, as mentioned above, limits the adjustments of landmarks. The purpose is to prevent deviations on the landmarks adjustments which would result in forming an allowable overall shape. With the model constraints, the final shape remains resemble to training shapes. This idea of global shape control is the main distinction between active contour model and active shape model. This shape model, similarly to profile model, is obtained across training shapes. The detail of the calculations and computational steps can be found in Cootes et al. [97].

In short, the final shape is determined by both the profile model and shape model: the profile model inspects landmarks surrounding and informs the landmarks about the movement suppose to make in order to fulfil as close as possible to the predefined profile; the shape model inspects the shape as a whole and informs landmark about the permissible movement while trying to fulfil the predefined profile.

ASM has been applied by Thodberg and Rosholm [30] to address the problem of hand bone segmentation. Extensive training that have to be done to complete the model in order to imitate the recognition understanding of human beings in segmenting the hand bone. Note that the initiation of set points placement to mark the spatial position of hand bone shape demands expert to be the operators. Both requirements of training set and human expert are the main weakness of this model in addressing the problem. It is tedious, subjective and time consuming to delineate the shape from a large training set, not to mention the critical issue of the availability of these resources. Therefore, an alternative segmentation framework has to be established when the resources are limited and this motivates the research of this book.

2.7.3 Active Appearance Model (AAM)

AAM is a statistical model of shape and grey-level appearance of the targeted object proposed by Edwards et al. [108]. The final aim is to generalize the model to all valid example [109]. The relationship between the model parameter displacements and the errors between training example and a model instance is learned during the training phase [110]. By computing the errors of fitting and using the previously obtained parameters, the current parameters with the intent of improving the current fitting can be updated.

AAM and the closely related to the concepts found in the methods of Active shape model. The AAMs is most frequently being adopted in the application related to face modeling. Besides face modeling [111], it has been implemented in other applications as well such as in medical image processing [112–114]. The typical first step of AAM is to fit the AAM to an input image using model parameters that maximize the matching criteria between the model instance and the input image. The model parameters are then passed to a classifier to yield t classification tasks.

Fitting the AAM to an input image involves solving a non-linear optimization problem. The conventional method of solving the problem is by updating the parameters iteratively. This update has to be incremental additive and the parameters refer to shape and appearance coefficients. The input image can be warped onto the model coordinate frame by using the current shape parameters estimations. The error between the model instance and the fitting of AAM onto the image can be computed. This error then acts as feedback in next iteration that would affect the updates of the parameters. The constant coefficients in this linear relationship between the updates and errors can then be found either by linear regression or by other numerical methods.

Although the AAM appears to be the useful model-based system in medical image segmentation, it has constraints that impede its performance in practical application [115].

1. Low efficiency in real-time systems: current algorithm of AAM consumes a lot of time and space computational costs. Thus, it is of prime importance to minimize the complexities in time and space needed to perform the algorithm in order to realize it in real-time system. The efficiency is mainly affected by the following factors: manual landmarks placement, complex texture representation in high resolution medical image, iterative procedure in solving the optimization problem.
2. Low discriminative ability for recognition and segmentation systems: only a group of object is being modeled and thus it is considered as a generative model which possesses no ability to classify different objects. This ability depends on the accuracy of model fitting which are affected by how the prior

shape is chosen; how the texture is represented; how the texture is modeled. It is crucial to improve this discriminative ability to perform segmentation tasks effectively.
3. Inconsistent robustness under different circumstances: the performance of the system is influenced by different conditions such as the existence of pose variations, uneven illumination, and absence of features, low resolution and the presence of noises.

AAM is a very useful model as it can capture the mode of variations of deformable objects given a set of training examples. The mode of variations includes shape and texture as a whole. Besides, it can perform the projection of object onto low dimensional subspace to reduce redundancy and capture main component of variations. Thus, it has been implemented in a lot of applications especially medical image segmentation. Nonetheless, it has limitation in efficiency, discriminative ability, and robustness in different condition. In the problem of hand bone segmentation, the same group of researchers that adopted ASM has extended their works by applying AAM [29, 116]. The weaknesses discussed in applying ASM remains because the technical differences between AAM and ASM enhance only the robustness in terms of prior knowledge and the information around the object that have been incorporated into the model, not the practicality in terms of availability of training set and expert operators.

The deformable models, ASM and AAM, undoubtedly are powerful segmentation methods. However, they are not without weaknesses and are not best method for automated hand segmentation. The reasons are summarized as follow:

1. The landmarks placement has to be manually annotated by users. Incorrect landmarks placements lead to unreliable capture of shape variability.
2. The number of landmarks has to be specified by user manually. Insufficient landmarks lead to failure in obtaining the shape of the targeted structures; excessive landmarks lead to computational inefficiency.
3. The training phase requires a lot of training examples in database which is not necessarily available in many applications. Insufficient training examples lead to failure in generalizing the mean structure's shape.
4. The nature of hand bone development of children: different number and size of bones in different range of age complicated the implementation ASM and AAM especially in establishing the general form of mean shape.
5. The alignment phase is uncertain in terms of its numerical stability: the convergence of the mean model in the iterative method has not been devised mathematically and prone to errors.
6. The choice in retaining the number of eigenvectors in principal component analysis has to be determined correctly by user. Incorrect decision leads to failure in capturing the representative points of the shape; consequently, inaccurate model is constructed and leads to undesired segmentation result.
7. Variations in hand structural positions are often largely deviated and this devotes to non-linear parameter relations that invariably impede the accurate segmentation as a whole.

2.8 Summary

In this chapter, the techniques of histogram equation, hand bone segmentation and image segmentation techniques have been reviewed. The pros and cons of those techniques have been analyzed and compared. It is obvious that those techniques are not sufficient to solve the hand bone segmentation problems. Hence, this motivates a new technique that capable of segmenting the hand bone problem to be established and implemented in the fully automated computer-aided skeletal age scoring system. The proposed novel framework of hand bone segmentation is presented in Chap. 3.

References

1. Aja-Fernández S, De Luis-García R, Martín-Fernández MÁ, Alberola-López C (2004) A computational TW3 classifier for skeletal maturity assessment. A Computing with Words approach. J Biomed Inform 37:99–107
2. Niemeijer M, Van Ginneken B, Maas CA, Beek FJA, and Viergever MA (2003) Assessing the skeletal age from a hand radiograph: Automating the tanner-whitehouse method. In: Sonka M and Fitzpatrick JM (eds), Proceedings of the 2003 SPIE Medical Imaging. vol 5032 II, pp 1197–205, San Diego, CA
3. Pietka E, Gertych A, Pospiech S, Cao F, Huang HK, Gilsanz V (2001) Computer-assisted bone age assessment: Image preprocessing and epiphyseal/metaphyseal ROI extraction. IEEE Trans Med Imaging 20:715–729
4. Pietka E, Kaabi L, Kuo ML, Huang HK (1993) Feature extraction in carpal-bone analysis. IEEE Trans Med Imaging 12:44–49
5. Somkantha K, Theera-Umpon N, Auephanwiriyakul S (2011) Boundary detection in medical images using edge following algorithm based on intensity gradient and texture gradient features. IEEE Trans Biomed Eng 58:567–573
6. Zhang J, Huang HK (1997) Automatic background recognition and removal (ABRR) in computed radiography images. IEEE Trans Med Imaging 16:762–771
7. Han CC, Lee CH, Peng WL (2007) Hand radiograph image segmentation using a coarse-to-fine strategy. Pattern Recogn 40:2994–3004
8. Hsieh CW, Jong TL, Tiu CM (2008) Carpal growth assessment based on fuzzy description. In Soft Computing in Industrial Applications, 2008. SMCia'08. IEEE Conference on 2008, Muroran, pp 355–358
9. Sotoca JM, Iñesta JM, Belmonte MA (2003) Hand bone segmentation in radioabsorptiometry images for computerised bone mass assessment. Comput Med Imaging Graph 27:459–467
10. Buie HR, Campbell GM, Klinck RJ, MacNeil JA, Boyd SK (2007) Automatic segmentation of cortical and trabecular compartments based on a dual threshold technique for in vivo micro-CT bone analysis. Bone 41:505–515
11. Smyth PP, Taylor CJ, Adams JE (1997) Automatic measurement of vertebral shape using active shape models. Image Vis Comput 15:575–581
12. Kass M, Witkin A, Terzopoulos D (1988) Snakes: active contour models. Int J Comput Vision 1:321–331
13. Mahmoodi S, Sharif BS, Chester EG, Owen JP, and Lee REJ (1997) Automated vision system for skeletal age assessment using knowledge based techniques. In: IEE international conference on image processing and its applications, 443 pt 2 (ed.), pp 809–13, IEE, Dublin, Irel
14. Mahmoodi S, Sharif BS, Graeme Chester E, Owen JP, Lee R (2000) Skeletal growth estimation using radiographic image processing and analysis. IEEE Trans Inf Technol Biomed 4:292–297

15. Sebastian TB, Tek H, Crisco JJ, Kimia BB (2003) Segmentation of carpal bones from CT images using skeletally coupled deformable models. Med Image Anal 7:21–45
16. Tristán-Vega A, Arribas JI (2008) A radius and ulna TW3 bone age assessment system. IEEE Trans Biomed Eng 55:1463–1476
17. Giordano D, Spampinato C, Scarciofalo G, Leonardi R (2010) An automatic system for skeletal bone age measurement by robust processing of carpal and epiphysial/metaphysial bones. IEEE Trans Instrum Meas 59:2539–2553
18. Jong-Min L, Whoi-Yul K (2008) Epiphyses extraction method using shape information for left hand radiography. In: Convergence and Hybrid Information Technology, 2008. ICHIT '08. International Conference on 28–30 Aug. 2008. pp 319–326
19. Michael DJ, Nelson AC (1989) HANDX: a model-based system for automatic segmentation of bones from digital hand radiographs. IEEE Trans Med Imaging 8:64–69
20. Manos G, Cairns AY, Ricketts IW, Sinclair D (1993) Automatic segmentation of hand-wrist radiographs. Image Vis Comput 11:100–111
21. Manos GK, Cairns AY, Rickets IW, Sinclair D (1994) Segmenting radiographs of the hand and wrist. Comput Methods Programs Biomed 43:227–237
22. Sharif BS, Zaroug SA, Chester EG, Owen JP, and Lee EJ (1994) Bone edge detection in hand radiographic images. In: Engineering in medicine and biology society, 1994. Engineering advances: new opportunities for biomedical engineers. Proceedings of the 16th annual international conference of the IEEE. pt 1 (ed.), vol 16, pp 514–5, Baltimore, MD, USA
23. Mahmoodi S, Sharif BS, Chester EG, Owen JP, and Lee REJ (1999) Bayesian estimation of growth age using shape and texture descriptors. In: Image processing and its applications, 1999. Seventh international conference on, 465 II (ed.), vol 2, pp 489–493, IEE, Manchester, UK
24. Pietka E, Pospiech-Kurkowska S, Gertych A, Cao F (2003) Integration of computer assisted bone age assessment with clinical PACS. Comput Med Imaging Graph 27:217–228
25. Pietka E, Gertych A, Pospiech-Kurkowska S, Cao F, Huang HK, Gilzanz V (2004) Computer-assisted bone age assessment: graphical user interface for image processing and comparison. J Digit Imaging 17:175–188
26. Hsieh CW, Jong TL, Chou YH, Tiu CM (2007) Computerized geometric features of carpal bone for bone age estimation. Chin Med J 120:767–770
27. Zhang A, Gertych A, Liu BJ (2007) Automatic bone age assessment for young children from newborn to 7-year-old using carpal bones. Comput Med Imaging Graph 31:299–310
28. Tran Thi My Hue MGS, Kim JY, Choi SH (2011) Hand bone image segmentation using watershed transform with multistage merging. J Korean Inst Inf Technol 9:59–66
29. Thodberg HH (2002) Hands-on experience with active appearance models. In: Sonka M, Michael Fitzpatrick J (eds), vol 4684 I, pp 495–506, San Diego, CA
30. Thodberg HH, Rosholm A (2003) Application of the active shape model in a commercial medical device for bone densitometry. Image Vis Comput 21:1155–1161
31. Thodberg HH, Kreiborg S, Juul A, Pedersen KD (2009) The BoneXpert method for automated determination of skeletal maturity. IEEE Trans Med Imaging 28:52–66
32. Sezgin M, Sankur B (2004) Survey over image thresholding techniques and quantitative performance evaluation. J Electron Imaging 13:146–168
33. Gonzalez R, Woods R (2007) Digital Image Processing, 3rd edn. Prentice Hall, Upper Saddle River
34. Bernsen J (1986) Dynamic Thresholding of Grey-Level Images. In: International conference on pattern recognition, pp. 1251–5, IEEE, Paris, France
35. Lee SU, Yoon Chung S, Park RH (1990) A comparative performance study of several global thresholding techniques for segmentation. Comput Vis, Graph, Image Process 52:171–190
36. Shapiro L and Stockman G (2001) Computer Vision. Prentice Hall, Upper Saddle River
37. Liyuan L, Ran G, Weinan C (1997) Gray level image thresholding based on fisher linear projection of two-dimensional histogram. Pattern Recogn 30:743–749
38. Baradez MO, McGuckin CP, Forraz N, Pettengell R, Hoppe A (2004) Robust and automated unimodal histogram thresholding and potential applications. Pattern Recogn 37:1131–1148

39. Yan F, Zhang H, Kube CR (2005) A multistage adaptive thresholding method. Pattern Recogn Lett 26:1183–1191
40. Tsai D-M (1995) A fast thresholding selection procedure for multimodal and unimodal histograms. Pattern Recogn Lett 16:653–666
41. Parker JR (1991) Gray level thresholding in badly illuminated images. IEEE Trans Pattern Anal Mach Intell 13:813–819
42. Zhao M, Yang Y, Yan H (2000) An adaptive thresholding method for binarization of blueprint images. Pattern Recogn Lett 21:927–943
43. Shafait F, Keysers D, and Breuel TM (2008) Efficient implementation of local adaptive thresholding techniques using integral images. In: Document recognition and retrieval XV, vol 6815, San Jose, CA
44. Huang Q, Gao W, Cai W (2005) Thresholding technique with adaptive window selection for uneven lighting image. Pattern Recogn Lett 26:801–808
45. Niblack W (1990) An Introduction to Digital Image Processing. Prentice-Hall, Upper Saddle River
46. De Santis A, Sinisgalli C (1999) A Bayesian approach to edge detection in noisy images. Circuits and systems i: fundamental theory and applications, IEEE Transactions on 1999, vol 46, pp 686–699
47. Whatmough RJ (1991) Automatic threshold selection from a histogram using the "exponential hull". CVGIP: Graph Models Image Process 53:592–600
48. Luijendijk H (1991) Automatic threshold selection using histograms based on the count of 4-connected regions. Pattern Recogn Lett 12:219–228
49. Guo R, Pandit SM (1998) Automatic threshold selection based on histogram modes and a discriminant criterion. Mach. Vision Appl 10:331–338
50. Otsu N (1979) A threshold selection method from gray-level histograms. IEEE Trans Syst, Man, Cybern 9:62–66
51. Lu C, Zhu P, Cao Y (2010) The segmentation algorithm of improvement a two-dimensional Otsu and application research. In: Software technology and engineering (ICSTE), 2010 2nd international conference on 2010. Vol. 1, pp V176–V179, San Juan, PR
52. Ge Y, Yang RF, Zhang P (2012) Research on the image segmentation method based on improved two-dimension Otsu arithmetic. Hedianzixue Yu Tance Jishu/Nucl Electron Detect Technol 32:112–115
53. Scharcanski J, Venetsanopoulos AN (1997) Edge detection of color images using directional operators. IEEE Trans Circuits Syst Video Technol 7:397–401
54. LingFei L, ZiLiang P (2008) An edge detection algorithm of image based on empirical mode decomposition. In Intelligent information technology application. IITA '08. Second international symposium on 2008. vol 1, pp 128–132
55. Bakalexis SA, Boutalis YS, and Mertzios BG (2002) Edge detection and image segmentation based on nonlinear anisotropic diffusion. In: Digital signal processing. DSP 2002. 2002 14th international conference on 2002. vol 2, pp 1203–6
56. Canny J (1986) A computational approach to edge detection. IEEE Trans Pattern Anal Mach Intell 8:679–698
57. You-yi Z, Ji-lai R, Lei W (2010) Edge detection methods in digital image processing. In: Computer science and education (ICCSE) 5th international conference on 2010. vol pp 471–473
58. Deriche R (1987) Using Canny's criteria to derive a recursively implemented optimal edge detector. Int J Comput Vis 1:167–187
59. Ding L, Goshtasby A (2001) On the Canny edge detector. Pattern Recogn 34:721–725
60. Duda RO, Hart PE (1972) Use of the Hough transformation to detect lines and curves in pictures. Commun ACM 15:11–15
61. Hough P (1962) Method and means for recognizing complex patterns. In: United States Patent Office

62. Chung CH, Cheng SC, Chang CC (2010) Adaptive image segmentation for region-based object retrieval using generalized Hough transform. Pattern Recogn 43:3219–3232

63. Xu X, Zhou Y, Cheng X, Song E, Li G (2012) Ultrasound intima–media segmentation using Hough transform and dual snake model. Comput Med Imaging Graph 36:248–258

64. Kassim AA, Tan T, Tan KH (1999) A comparative study of efficient generalised Hough transform techniques. Image Vis Comput 17:737–748

65. Shapiro V A (1996) On the hough transform of multi-level pictures. Pattern Recogn 29:589–602

66. Hart PE (2009) How the Hough transform was invented [DSP History]. IEEE Signal Processing Magazine 26:18–22

67. Ji J, Chen G, Sun L (2011) A novel Hough transform method for line detection by enhancing accumulator array. Pattern Recogn Lett 32:1503–1510

68. Zheng L, Shi D (2011) Advanced Radon transform using generalized interpolated Fourier method for straight line detection. Comput Vis Image Underst 115:152–160

69. Illingworth J, Kittler J (1987) The Adaptive Hough Transform. IEEE Trans Pattern Anal Mach Intell 9:690–698

70. Nixon M (1990) Improving an extended version of the Hough transform. Signal Processing 19:321–335

71. Kiryati N, Eldar Y, Bruckstein AM (1991) A probabilistic Hough transform. Pattern recognition 24:303–316

72. Hanif T, Sandler MB (1994) A counter-based Hough transform system. Microprocess Microsyst 18:19–26

73. Torii A, Imiya A (2007) The randomized-Hough-transform-based method for great-circle detection on sphere. Pattern Recogn Lett 28:1186–1192

74. Ballard DH (1981) Generalizing the Hough transform to detect arbitrary shapes. Pattern Recogn 13:111–122

75. Lo RC, Tsai WH (1995) Gray-scale hough transform for thick line detection in gray-scale images. Pattern Recogn 28:647–661

76. Kang CC, Wang WJ, Kang CH (2012) Image segmentation with complicated background by using seeded region growing. AEU—International Journal of Electronics and Communications 66:767–771

77. Fan J, Zeng G, Body M, Hacid M-S (2005) Seeded region growing: an extensive and comparative study. Pattern Recogn Lett 26:1139–1156

78. Grinias I, Tziritas G (2001) A semi-automatic seeded region growing algorithm for video object localization and tracking. Sig Process: Image Commun 16:977–986

79. Lin GC, Wang WJ, Kang CC, Wang CM (2012) Multispectral MR images segmentation based on fuzzy knowledge and modified seeded region growing. Magn Reson Imaging 30:230–246

80. Mehnert A, Jackway P (1997) An improved seeded region growing algorithm. Pattern Recogn Lett 18:1065–1071

81. Tremeau A, Borel N (1997) A region growing and merging algorithm to color segmentation. Pattern Recogn 30:1191–1203

82. Digabel H, Lantuéjou C (1978) Iterative algorithms. In: Actes du Second Symposium Europeen d'Analyse Quantitative des Microstructures en Sciences des Materiaux, Biologie et Medecine. pp 4–7 Caen

83. Beucher S and Lantuejoul C (1979) Use of Watersheds in Contour Detection. In: International workshop on image processing: Real-time edge and motion detection/estimation, Rennes, France

84. Vincent L, Soille P (1991) Watersheds in digital spaces: an efficient algorithm based on immersion simulations. IEEE Trans Pattern Anal Mach Intell 13:583–598

85. Cousty J, Bertrand G, Najman L, Couprie M (2010) Watershed cuts: thinnings, shortest path forests, and topological watersheds. IEEE Trans Pattern Anal Mach Intell 32:925–939

86. Haris K, Efstratiadis SN, Maglaveras N, Katsaggelos AK (1998) Hybrid image segmentation using watersheds and fast region merging. IEEE Trans Image Process 7:1684–1699
87. Kim JB, Kim HJ (2003) Multiresolution-based watersheds for efficient image segmentation. Pattern Recogn Lett 24:473–488
88. Frucci M, Ramella G, Sanniti di Baja G (2007) Using resolution pyramids for watershed image segmentation. Image Vis Comput 25:1021–1031
89. Jung CR, Scharcanski J (2005) Robust watershed segmentation using wavelets. Image Vis Comput 23:661–669
90. Jung CR (2007) Combining wavelets and watersheds for robust multiscale image segmentation. Image Vis Comput 25:24–33
91. Hamarneh G, Li X (2009) Watershed segmentation using prior shape and appearance knowledge. Image Vis Comput 27:59–68
92. Masoumi H, Behrad A, Pourmina MA, Roosta A (2012) Automatic liver segmentation in MRI images using an iterative watershed algorithm and artificial neural network. Biomed Signal Process Control 7:429–437
93. Meyer F (1994) Topographic distance and watershed lines. Signal Processing 38:113–125
94. Tarabalka Y, Chanussot J, Benediktsson JA (2010) Segmentation and classification of hyperspectral images using watershed transformation. Pattern Recogn 43:2367–2379
95. McInerney T, Terzopoulos D (2000) Deformable models. Academic Press, San Diego
96. Leymarie FF (1990) Tracking and Describing Deformable Objects Using Active Contour Models [Thesis]
97. Cootes TF, Taylor CJ, Cooper DH, Graham J (1995) Active Shape Models-Their Training and Application. Comput Vis Image Underst 61:38–59
98. Xue Z, Li SZ, Teoh EK (2003) Bayesian shape model for facial feature extraction and recognition. Pattern Recogn 36:2819–2833
99. Zheng Z, Jiong J, Chunjiang D, Liu X, Yang J (2008) Facial feature localization based on an improved active shape model. Inf Sci 178:2215–2223
100. Sukno FM, Guerrero JJ, Frangi AF (2010) Projective active shape models for pose-variant image analysis of quasi-planar objects: Application to facial analysis. Pattern Recogn 43:835–849
101. Jang D-S, Choi H-I (2000) Active models for tracking moving objects. Pattern Recogn 33:1135–1146
102. Kim W, Lee J–J (2005) Object tracking based on the modular active shape model. Mechatronics 15:371–402
103. Nuevo J, Bergasa LM, Llorca DF, Ocaña M (2011) Face tracking with automatic model construction. Image Vis Comput 29:209–218
104. Liu Z, Shen H, Feng G, Hu D (2012) Tracking objects using shape context matching. Neurocomputing 83:47–55
105. Hodge AC, Fenster A, Downey DB, Ladak HM (2006) Prostate boundary segmentation from ultrasound images using 2D active shape models: optimisation and extension to 3D. Comput Methods Programs Biomed 84:99–113
106. Aung MSH, Goulermas JY, Stanschus S, Hamdy S, Power M (2010) Automated anatomical demarcation using an active shape model for videofluoroscopic analysis in swallowing. Med Eng Phys 32:1170–1179
107. Toth R, Tiwari P, Rosen M, Reed G, Kurhanewicz J, Kalyanpur A et al (2011) A magnetic resonance spectroscopy driven initialization scheme for active shape model based prostate segmentation. Med Image Anal 15:214–225
108. Edwards GJ, Taylor CJ, Cootes TF (1998) Interpreting face images using active appearance models. In FG '98: Proceedings of the 3rd international conference on face & gesture recognition. IEEE computer society
109. Cootes TF, Page GJ, Jackson CB, Taylor CJ (1996) Statistical grey-level models for object location and identification. Image Vis Comput 14:533–540

110. Cootes TF, Edwards GJ, Taylor CJ (2001) Active Appearance Models. IEEE Trans Pattern Anal Mach Intell 23:681–685
111. Butakoff C, Frangi AF (2010) Multi-view face segmentation using fusion of statistical shape and appearance models. Comput Vis Image Underst 114:311–321
112. Roberts M, Cootes T, Pacheco E, Adams J (2007) Quantitative Vertebral Fracture Detection on DXA Images Using Shape and Appearance Models. Academic Radiology 14:1166–1178
113. Andreopoulos A, Tsotsos JK (2008) Efficient and generalizable statistical models of shape and appearance for analysis of cardiac MRI. Med Image Anal 12:335–357
114. Patenaude B, Smith SM, Kennedy DN, Jenkinson M (2011) A Bayesian model of shape and appearance for subcortical brain segmentation. NeuroImage 56:907–922
115. Gao X, Su Y, Li X, Tao D (2010) A Review of Active Appearance Models. Systems, Man, and Cybernetics, Part C: Applications and Reviews, IEEE Transactions on 40:145–158
116. Thodberg HH, Van Rijn RR, Tanaka T, Martin DD, Kreiborg S (2010) A paediatric bone index derived by automated radiogrammetry. Osteoporos Int 21:1391–1400

Chapter 3
Design and Implementation

Abstract In previous chapter, the weaknesses of conventional segmentation methods have been identified. This concludes the desired segmentation criteria in order to guide the mechanism of the proposed framework of segmentation. The segmentation is performed to partition the hand bone from its background and soft-tissue region in the beginning of this chapter. The challenges of hand bone segmentation are the overlapping intensity between the soft-tissue region and the spongy bone region within the hand bone. Three modules of techniques will be discussed and implemented to solve the problem in this chapter.

3.1 Introduction

Top-down strategy is adopted in designing the proposed segmentation framework. Firstly, the overview of the desired system is obtained through literature reviews by reviewing the existing techniques and analyzing the factors leading them to failure as effective hand bone segmentation technique. After gaining some insights on constituting a desired hand bone segmentation framework, the emphasis is on identifying the desired characteristics, only then each sub-framework to satisfy each requirement is proposed.

The desired characteristics of hand bone segmentation framework adopted in computer-aided skeletal age scoring system should comprise the follow:

1. Contrast, illumination, orientation invariants: to improve consistent segmentation robustness under different conditions of X-ray settings and devices.
2. Relatively low computational complexity: to realize practical execution time for automated BAA system. Ideally, it is comprehensive enough to tackle with image complexities and uncertainties yet it is simple enough to be executed in a reasonable timeframe.

Y. C. Hum, *Segmentation of Hand Bone for Bone Age Assessment*, 47
SpringerBriefs in Applied Sciences and Technology, DOI: 10.1007/978-981-4451-66-6_3,
© The Author(s) 2013

3. No complicated 'training' procedures: to prevent high dependency on availability of training samples of hand bone radiographs. However, simple parameter tuning procedures without depending on availability of training hand bone radiographs have to be established to capture the variations of uncertainties in image nature.

4. Utilization of prior knowledge: to ensure the usage of available information to optimize the result on hand bone segmentation. Besides, making use of 'by-products' of image pre-processing is preferable.

5. Relatively high resistance to noises: to improve performance of segmentation despite the inevitable random signals in the hand bone radiographs.

6. Automated or minimum dependency on human interventions: to ensure objectivity, to enable reproducibility and to avoid laboriousness.

7. Consistent accuracy: to ensure relatively high precision in segmentation on resultant hand bone for subsequent processing in automated BAA system.

8. Resemblance to manual segmentation: to ensure a certain level of artificial intelligence in the designed algorithm to emulate human visual perception.

9. No overdependence on certain image feature: to enhance segmentation robustness under absence of any certain property such as intensity discontinuity or edges.

10. Adaptability: to increase segmentation robustness under presence of variability in different regions of hand radiographs.

11. Optimality: all parameters are chosen based on the direction of finding the optimum solution and not arbitrarily pre-set. However, this criterion should not violate the second criterion.

To facilitate the subsequent explanations on the proposed framework, henceforth, aforementioned desired property is referred as P1, P2, P3 … and so forth. For example, the first property of contrast, illumination, orientation invariance is referred as P1 and the tenth criterion of adaptability is referred as P10.

3.2 The Proposed Segmentation Framework

The proposed segmentation is a series of image processing procedures that is specially tailored for solving the problem of hand bone segmentation. The procedure begins with pre-processing to improve the input image properties so that it is ready to go for the central processing algorithm namely the Adaptive Crossed Reconstructed (ACR) algorithm. After that, it will undergo a feedback process to assure the quality of the segmented image. The flowchart of the main procedures is shown in Fig. 3.1.

Pre-processing, main processing algorithm and quality assurance process represent the subroutines of the proposed framework. The details of each subroutine are discussed follow.

Fig. 3.1 The framework of
the proposed segmentation
framework

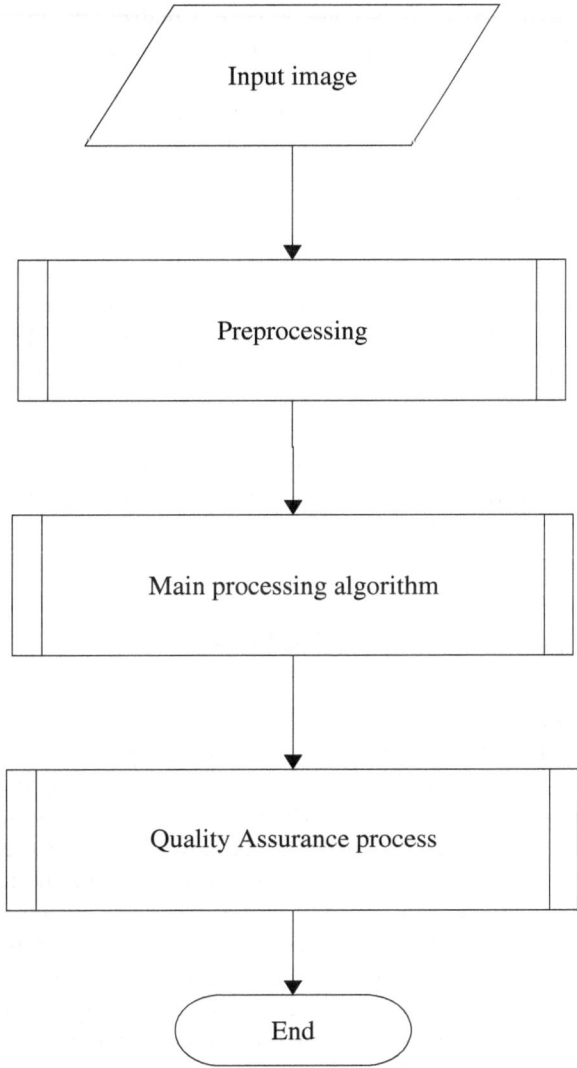

3.3 Pre-processing

Before adopting the proposed Adaptive Crossed Reconstruction (ACR) segmentation algorithm, the input image has to undergo two pre-processing procedures to serve two purposes: (1) to equalize the brightness of the image (2) to decrease the variations in brightness different in bone structure. The first purpose is realized by using a proposed histogram equalization specially designed for hand bone pre-processing namely the Multipurpose Beta Optimized Bi-Histogram Equalization

(MBOBHE). The second purpose is realized by adopting the anisotropic diffusion technique which is conventionally adopted in ultrasound image noise reduction process.

The first aim of equalizing the brightness is to standardize the image and this in turn increases the robustness of the proposed procedure to various hand bone radiographs from different X-ray machine with different range of intensity distribution due to environmental factor or the devices internal factor. The second aim of equalizing the brightness is to enhance the contrast of the input image to ease the detection of the pertinent features. Both aims prepare the input image to improve the result of subsequent procedures. The technique adopted is histogram equalization with some modifications and improvements. Firstly, the review on histogram equalization technique is conducted to assure that the current histogram equalization methods are unable to cope with the inherent problems of hand bone radiograph contrast enhancement. Then, modified histogram equalization is proposed and designed to optimize the visual effect of resultant equalized image for multiple objectives. The details of the designed histogram equalization method are discussed in next section.

3.3.1 The Proposed MBOBHE

The contrast of bone texture in ossification sites is of critical importance in the bone age assessment especially for TW3 method. An optimum contrast adjusted ossification site image can ease the subsequent procedures in computer-aided skeletal age scoring system. The first advantage is that the outline of the bone structure can be perceived clearly. Second advantage is devoted to ossification development stage evaluation procedures by making both the structure and texture of ossification bones to be well-defined and the important features are visually and computationally distinguishable. Hence, a good contrast as pre-processing in computer-aided skeletal age scoring system is prominent for accurate determination of bone age.

Therefore, the objective is to find an optimal solution to enhance the contrast, the brightness and detail preservation of output image simultaneously. A new histogram equalization method is proposed, named Multipurpose Beta Optimal Bi-histogram Equalization (MBOBHE). This method disagrees with traditional improvements of histogram equalization that tend to be optimized by minimizing or maximizing only one of the properties such as contrast, brightness preservation and detail preservation. Instead, a histogram equalization that capable of optimizing all the properties simultaneously by using regularization function rather than by minimization or maximization any single objective such as brightness preservation, contrast enhancement or detail preservation is proposed. The motivation of the design is due to the fact which we will prove qualitatively in this book later, that a visually good contrast image should be optimized in terms of holistic performance of all properties. Holistic performance that takes all properties into account can produce more natural output image.

The proposed MBOBHE can be perceived as simultaneous accomplishments of three objectives functions using Pareto optimization of bi-modal model [1]. The aim is to improve the contrast without producing the mean-shift and detail loss artifacts. In terms of Pareto optimization, this improvement without sacrificing the others is regarded as Pareto-efficiency or Pareto-optimality, whereas the contrast, brightness mean preservation and details preservations are regarded as Pareto criteria. MBOBHE is different from the previous MMBEBHE that performs only a single objective function optimization of bi-modal model as in [2]. The main motivation of proposing MBOBHE is to design a HE that can

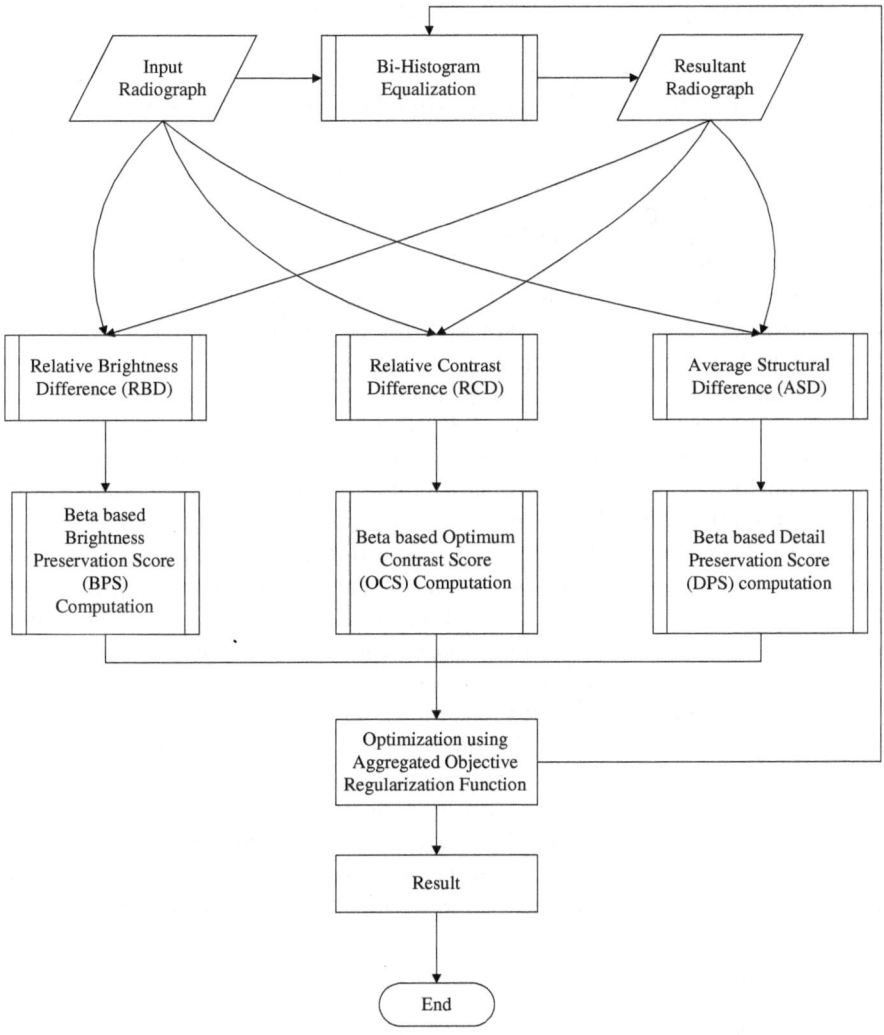

Fig. 3.2 The flowchart of the proposed MBOBHE

generate natural contrast enhanced resultant image with least artifacts such as mean-shift and detail loss by searching an optimal point to perform the bi-modal and taking three objectives into considerations: contrast enhancement, brightness preservation and details preservation.

In this section, the design of the proposed algorithm will be discussed. Generally, there are four main operations involved in MBOBHE: modeling of each criterion as single modal objective beta function, optimal solution of the aggregated multiple objectives function, histogram decomposition into two sub-histograms using computed Pareto solution, and execution of GHE on each sub-histogram. The overall framework of the proposed MBOBHE is shown in Fig. 3.2 and the details of these steps are explained in accordance to the four main operations in Sects. 3.3.1.1, 3.3.1.2, 3.3.1.3 and 3.3.1.4.

3.3.1.1 Modeling of Criteria as Single Modal Objective Beta Function

To construct the regularization function for multiple objective histogram equalization, the image quality metric used to gauge good qualities output image from poor qualities output image must be first identified as they are the fundamental components in the objective function [3]. Similar to MMBEBHE, the optimization of brightness preservation is based on the image quality metric; Unlike MMBEBHE, Average Mean Brightness Error (AMBE) is not adopted, and the proposed method has multiple objectives.

It is a convention to adopt Mean Squared Error (MSE), and the converted form of MSE (Peak Signal-to-Noise Ratio) PSNR due to its simplicity, high intuitiveness and parameter-free characteristic. This is true despite the fact that MSE as a signal fidelity measures, demonstrates fallible performance and exhibits severe drawbacks, particularly when it involves perceptually significant signals such as images. The PSNR is useful if images contain different dynamic ranges to be compared, but otherwise, it contains no new information relative to the MSE. Besides, the output values are not within a bound set of values and it possesses no unique maxima [4]. Furthermore, none of the abovementioned quality metrics could accurately measure the structural changes in an image.

On the ground of the above mentioned drawbacks of conventional performance metrics, a new framework of quality metrics that capable of quantitatively modeling the brightness preservation, contrast and detail change after the histogram equalization operation is designed. The three metrics or performance criteria are termed as Brightness Preservation Score (BPS) function, Optimum Contrast Score (OCS) function, and Detail Preservation Score (DPS) function. Each metric is then combined together to form the complete model of the regularization function which is the gist of the MBOBHE.

(A) The Brightness Preservation Score (BPS) Function

The *BPS* is designed as a function output which is a numerical score that represents the ability of brightness preservation, corresponds to the brightness difference between the input image and the histogram equalized image. The function is designed such that the Relative Brightness Difference (RBD) are the function domain, is the parameter, and the resultant values are bound between the range of 0 and 1 to enable comparisons. The step-by-step construction details of BPS function are discussed as follows:

$$RBD = \frac{|\mu_y - \mu_x|}{\mu_x + c} \tag{3.1}$$

The RBD denotes the relative brightness difference between the input radiograph and histogram equalized radiograph, denotes input image, denotes output image, denotes mean of input image gray level intensity, denotes mean of output image gray level intensity. It is too 'raw' and not yet suitable to be used solely as component in the multiple objective regularization function as it lacks of two desired important features: The first feature is the standard bound function range value which is critical in the final regularization function to establish an integrated performance from all the components; it is obvious to notice that the output value can be as large as infinity in extreme cases when the brightness mean of input radiograph is 0.

$$BPS\,(RBD;m,n) = \frac{1}{\beta\,(m,n)} RBD^{(m-1)}(1 - RBD)^{(n-1)} \tag{3.2}$$

$$NBPS\,(RBD;m,n) = \frac{BPS(RBD;m,n)}{argmax(BPS\,(RBD;m,n))} \tag{3.3}$$

The Eq. (3.2) denotes the Brightness Preserving Score (BPS) function in the context of histogram equalization which bears a resemblance to Beta probability distribution function where RBD satisfies with two real positive parameters a and b manipulating the relative significance of initial brightness difference between output and input image. Quantity denotes Beta function in terms of gamma function:

$$\beta\,(m,n) = \frac{\Gamma\,(m)\,\Gamma(n)}{\Gamma(m+n)} \tag{3.4}$$

$$\text{where } \Gamma\,(a) = \int_0^1 e^{-t}t^{(a-1)}dt \tag{3.5}$$

Equation (3.3) denotes the Normalized Brightness Preserving Score (NBPS). The effect of both parameters m and n are illustrated in Figs. 3.3 and 3.4.

What is the difference between NBPS and BPS? What BPS indicates? The difference between NBPS and BPS is that the NBPS assures that the metric is bound between 0 and 1. As in Fig. 3.5, each BPS function has different maximum value and the maximum value is not bound. However, as illustrated in Fig. 3.6, the maximum values of all the NBPS functions with different parameters are identical, which is 1. The BPS provides us an index to measure quantitatively the ability of brightness preservation of the histogram equalized algorithm. The NBPS is

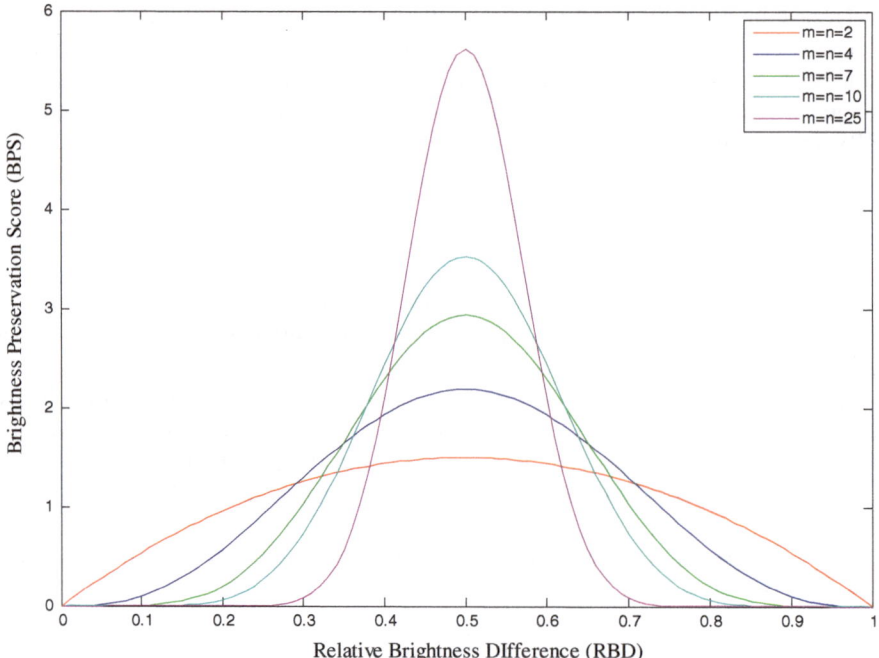

Fig. 3.3 Illustration of the relationship between *RBD* and *BPS* when both the shape parameters are equivalent: symmetrical with a central peak at 0.5 of *RBD*. The five curves in the figure demonstrate the effect of five different values of equivalent shape parameters on the shape of *BPS* function: as both shapes parameters increase, the curve spread and falls off evenly

normalized *BPS* that produces bound output values between 0 and 1; this bound output property is of prime importance to compare other measurements which will be defined later.

The *NBPS* of value 1 indicates perfect brightness preservation, whereas *NBPS* value close to 0 indicates poor brightness preservation. In other words, the more the *NBPS* approaches 1, the more perfect the brightness preservation. But how large is considered as 'large'? This ambiguity is explained by m and n; different values of m and n define different values of *RBD* required for favorable brightness preservation. For example, as illustrated in the Fig. 3.6, the red line, which is formed by using m = 2, n = 3 demonstrates that, for instance, if the *RBD* falls near 3.3, the histogram equalization has better performance in brightness preserving than histogram equalization algorithm that produces *RBD* around 3.2 or 3.8. As the figure illustrates, the *RBD* is around 3.3, compared to *RBD* around 3.2 or 3.8, produces higher *NBPS*. As shown in the Figs. 3.4 and 3.6, different values of m and n determine the value of *RBD* should have in order to produce high *NBPS*. In other words, the *NBPS* function regularizes the definition of 'good' or 'bad' brightness preserving ability, in terms of *RBD*.

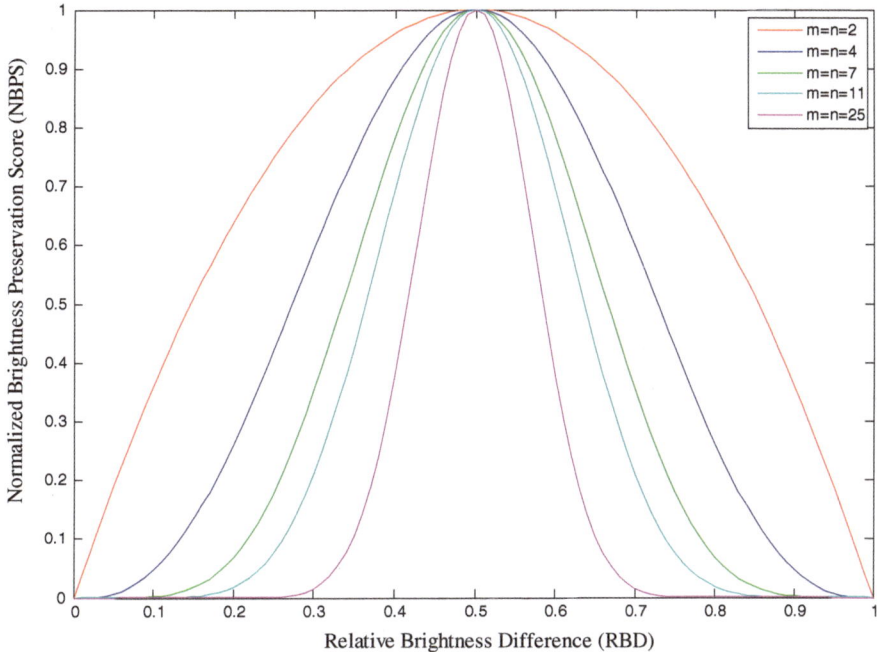

Fig. 3.4 Illustration of the bound property of *NBPS* function range when the *BPS* curves in BPS are normalized

One important feature of the proposed *BPS* is that it has revolutionized the traditional perception for brightness preservation which associates with minimizing the brightness difference to better brightness preserving ability, which is incorrect in the context of contrast enhancement on the ground that small brightness difference might indicate negligible contrast enhancement and this in turn will obscure the pertinent information. Therefore, the proposed new concept is to optimize the brightness different for better brightness preserving ability along with two important properties: contrast enhancement and detail preservation. This regularization function provides a new insight and technique in extending the flexibility of histogram equalization in brightness preservation so that it can be applied in various types of images. The next section will discuss about the proposed metric used to measure the contrast enhancement. This metric will finally be adopted in the final regularization objective function.

(B) The Optimum Contrast Score (OCS) Function

The Optimum Contrast Score (OCS) is designed as a function that output a numerical score which can be used to gauge the contrast enhancement corresponds to the contrast change between the input image and the histogram equalized image. The function is designed so that the contrast change is the function domain, p and q are the parameters, and the bound values between 0 and 1 are the range. The detail of the function is discussed as follows:

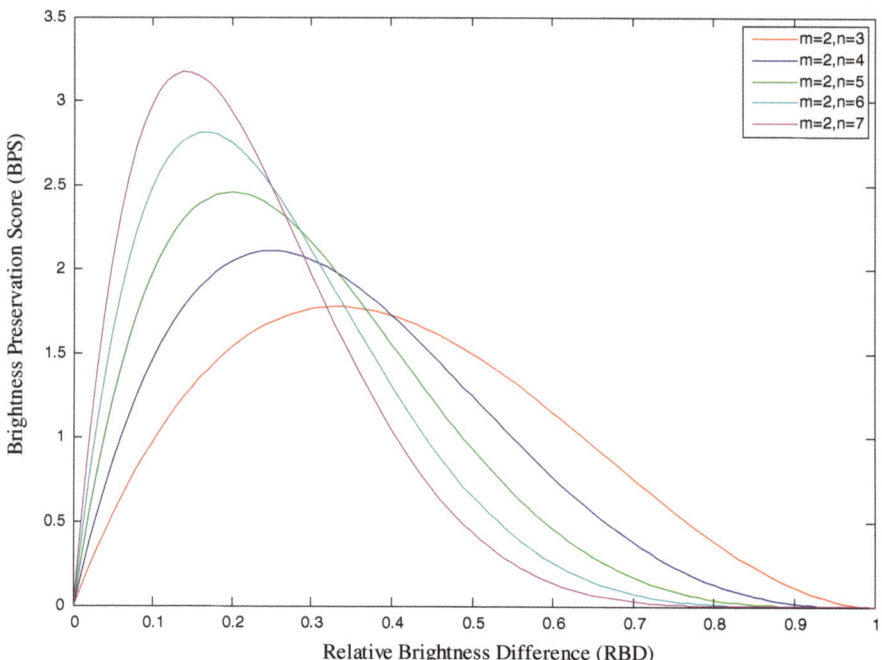

Fig. 3.5 Illustration of the relationship between *RBD* and *BPS* when both the shape parameters are not equivalent: asymmetrical triangular shaped distribution with peak at different value of *RBD*. These five curves demonstrate the effect of five different values of manipulated parameter n with constant parameter m = 5 on the shape of *BPS* function: as both shape parameters increase, the curve spread and falls off unevenly

$$RCD = \frac{1}{\frac{|\sigma_y - \sigma_x|}{\sigma_{x+c}}} + c \tag{3.6}$$

where
$$\sigma_y = \sigma_x \text{ if } \sigma_y > \sigma_x \tag{3.7}$$

$$\sigma_x = \sqrt{\frac{1}{MN}\sum_{i=0}^{N-1}\sum_{j=0}^{M-1}\left(\frac{x_{i,j}}{x_{max}} - \mu_x\right)^2} \tag{3.8}$$

$$\sigma_y = \sqrt{\frac{1}{MN}\sum_{i=0}^{N-1}\sum_{j=0}^{M-1}\left(\frac{x_{i,j}}{x_{max}} - \mu_y\right)^2} \tag{3.9}$$

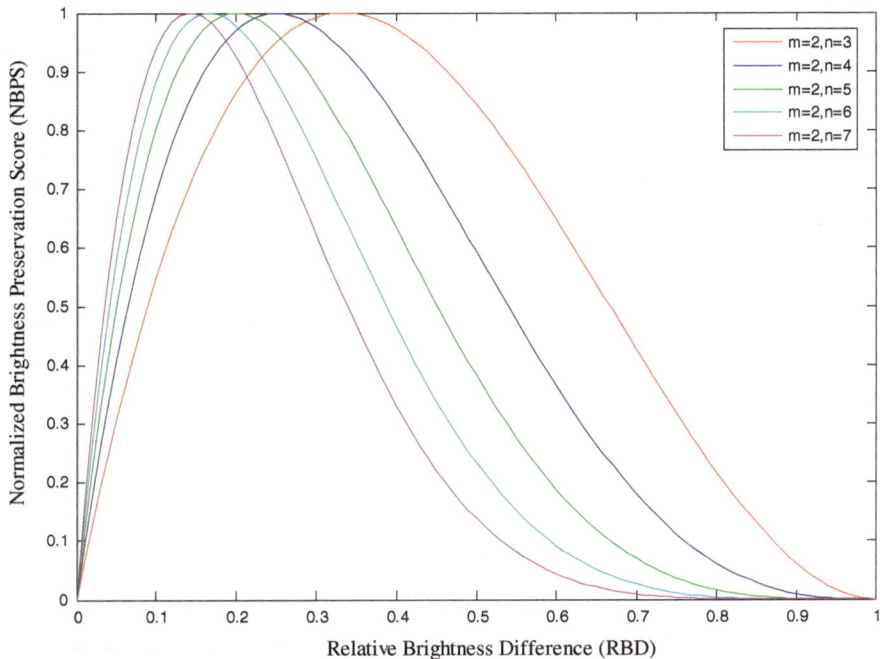

Fig. 3.6 Illustration of the bound property of *NBPS* function range when the *BPS* curves in Fig. 3.5 are normalized

Firstly, we define the input of *OCS* which is the Relative Contrast Different (RCD) as follows: where σ_x and σ_y denote Normalized Root Mean Squared (NRMS) contrast of input image where pixel intensities are the i-th j-th element of the two dimensional image of size by *M* by *N*, denotes an extremely small constant for purpose of maintaining the function stability. The *RCD* function represents a contrast comparison metric where the constant c is added for the sake of function stability. This function out satisfies bound condition where the range is limited between zero and one; function output is zero only when both input and output images possess identical standard deviation; function output is one whenever the output image NRMS is twice or more than NRMS of input image. Similar to *BPS*, the *RCD* is converted to *OCS* and *NOCS* using beta distribution function:

$$OCS\,(RCD; p, q) = \frac{1}{\beta\,(p, q)} RCD^{(p-1)}(1 - RCD)^{(q-1)} \qquad (3.10)$$

$$NOCS\,(p, q) = \frac{OCS(p, q)}{argmax(OCS\,(p, q))} \qquad (3.11)$$

This is to satisfy the new perception that distinguishable features in image of any application should have optimum contrast which is oppose to the traditional perception that the value of contrast should be as high as possible. The property of triangular shaped distribution with correctly tuned parameters satisfies this new perception. The new perception is based on the observation that relatively low contrast and relatively high contrast are both noticeably inferior to the resultant image; conversely, the optimum contrast is favorable. The reason is that relatively low contrast will obscure the pertinent features image which is significant in application; contrarily, relatively high contrast will produce over-enhanced contrast and lead to the saturated intensity artifacts.

(C) The Detail Preservation Score (DPS) Function

Firstly, we define the input of DPS which is the Average Structural Different (ASD) as follows:

$$d\left(X, Y\right) = \left(x_{i,j} - \mu_x\right) \left(y_{i,j} - \mu_y\right) \tag{3.12}$$

$$ASD\left(X, Y\right) = \frac{1}{MN} \sum_{i=0}^{N} s(x_{i,j}, y_{i,j}) \begin{cases} 1 \text{ if } d(x_{i,j}, y_{i,j}) \geq 0 \\ 0 \text{ if } d(x_{i,j}, y_{i,j}) \leq 0 \end{cases} \tag{3.13}$$

The Detail Preservation Score (DPS) is designed as a function that output a numerical score which can be used to gauge the performance of the detail preservation ability of the histogram equalization corresponds to the simultaneous discrepancy of pixel-mean intensity between the input image and the histogram equalized image. The function is designed so that the simultaneous pixel-mean intensity change is the function domain, U and V are the parameters, and the bound values between 0 and 1 are the function range. The detail of the function is discussed as follows:

3.3.1.2 Optimal Solution of the Aggregated Multiple Objectives Function

Solving the multiple objectives optimization amount to searching an optimal n-dimensional variables vector, $\vec{x} = \{x_1 \ldots x_n\}$ which is known as the 'optimization parameters' that fulfil all imposed constraints and at the same time optimizes all the conflicting performance criteria or objectives, represented by a m-dimensional vector objective function $\vec{f}(\vec{x}) = [f_1 \ldots f_m]$ [5]. In the context of multiple objectives histogram equalization, there is only one optimization parameter (n = 1), which is the decomposition point of bi-modal histogram equalization, and there are 3 objectives (m = 3) which are the maximum contrast, minimum brightness mean-shift and minimum detail loss. As mentioned, minimizing the mean-shift or detail loss is conflicting with maximizing the enhanced contrast; therefore, we need to find an optimal feasible (or Pareto optimal) decomposition point that can satisfy the three objectives simultaneously; the set of all Pareto optimal solution is called the efficient frontier.

The optimal solution is obtained via optimization of aggregated three performance criteria functions using weighted-sum approach [6]. Three criteria are summed up into a single objective function and each criterion is given a coefficient, to define the relative importance of each criterion in the aggregated objective function as formulated as followings:

$$F(x) = \sum_{i=1}^{N} c_i f_i(x)$$
(3.14)

$$where f_i = \{NBPS, NOCS, NDPS\}$$

Correctly tuned coefficients are essential to provide correct relative correlation between the measurements in *RBD*, *RCD* and *ASD* to the corresponding scores in *NBPS*, *NOCS* and *NDPS* respectively. These coefficients settings do not have standard rules of setting, it depends on the applications and the quality of the input images [7]. The added value of histogram equalization is that this technique standardizes the intensity distribution of the image luminance. Therefore, invariably, the parameters setting MBOBHE depends heavily on the applications. In the context of hand bone segmentation application, these coefficients are set to have equal importance. The optimization is then solved by evolutionary algorithm [8]. In next chapter of result and analysis, the performance of MBOBHE in the context of BAA using TW3 method will be justified via qualitative analysis by benchmarking the MBOBHE to existing types of HE methods such as GHE, BBHE, DSIHE, MMBEBHE, RMSHE and RSIHE.

3.3.1.3 Histogram Decomposition

The computed Pareto solution is used as a decomposing point to decompose the histogram into two sub-images: a lower sub-image and upper sub-image represented by X_L and X_U respectively based on the decomposing point, this process can be described using the mathematical Eq. (3.15).

$$X = X_L \cup X_U$$
(3.15)

where

$$X_L = \{X(i,j) | X(i,j) \leq X_S, \forall X(i,j) \in X\}$$

$$X_U = \{X(i,j) | X(i,j) > X_S, \forall X(i,j) \in X\}$$

In other words, the lower sub-image X_L is constituted of $\{X_0, X_1, \ldots, X_S\}$ and the upper sub-image X_U is constituted of $\{X_{S+1}, X_{S+2}, X_{S+3}, \ldots, X_{L-1}\}$. Suppose n_L^k and n_U^k represent the number of pixels with gray level X_k in X_L and X_U respectively, n_L denotes the total number of pixels in sub-image X_L, as expressed in Eq. (3.16); n_L denotes the total number of pixels in sub-image X_U, as expressed in Eq. (3.17). The probability density function (PDF) of sub image X_L is defined as in Eq. (3.18) and the *PDF* of sub-image X_U is defined as in Eq. (3.19)

$$n_L = \sum_{k=0}^{S} n_L^k \tag{3.16}$$

$$n_L = \sum_{k=S+1}^{L-1} n_U^k \tag{3.17}$$

$$P_L\,(X_k) = \frac{n_L^k}{n_L}$$
where $k = 1,2\ldots S$ \hfill (3.18)

$$P_U\,(X_k) = \frac{n_U^k}{n_U}$$
where $k = S+1, S+2,\ldots,L-1$ \hfill (3.19)

3.3.1.4 Execution of GHE on Each Sub-Histogram

After decomposing the input histogram into two sub-histograms, each sub-histogram undergoes independent GHE. Let n depicts the total number of the input image, defined as Eq. (3.20). The Cumulative Density Function (CDF) for X_L and X_U are represented as Eqs. (3.21) and (3.22), respectively.

$$n = n_L + n_U \tag{3.20}$$

$$c_L(X_k) = \sum_{k=0}^{S} P_L(X_k) \tag{3.21}$$

$$c_U(X_k) = \sum_{k=S+1}^{L-1} P_U(X_k) \tag{3.22}$$

By definition, it is expected that $c_L\,(X_{L-1}) = 1$. Once the histogram partition has completed, the conventional HE is implemented to each sub-histogram by utilizing the defined CDF as a transform function, and subsequently it is combined to form the output, Y, which can be expressed mathematically as follows:

$$Y_L\,(X_k) = X_0 + (X_S + X_0)\,C_L(X_k)$$
where $k = 0,1,\ldots, S$ \hfill (3.23)

$$Y_U\,(X_k) = X_{S+1} + (X_{L-1} + X_{S+1})\,C_U(X_k)$$
where $k = S+1, S+2,\ldots,L-1$ \hfill (3.24)

$$Y = Y_L + Y_U \tag{3.25}$$

The histogram equalization operation has remapped the gray level of the sub-image X_L over the range of (X_0, X_L) whereas Y_U has remapped the gray level of the sub-image X_U, over the range of (X_{S+1}, X_{L-1}). In other words, the grey level on input image has been remapped to the entire dynamic range in the output image.

3.3.2 The Application of Anisotropic Diffusion

Before implementing image processing techniques such as segmentation and pattern recognition, a prior filtering is used to decrease the level of noise in the input radiograph. There is, nonetheless, an inevitable problem associated with the conventional linear filtering such as Gaussian filtering: as the noise is being diffused, the boundaries are also diffused simultaneously. The diffusion of noise of first condition is desirable; but the second condition of boundary smoothing is inferior to segmentation. To avoid this undesired diffusion, a non-linear anisotropic diffusion method are constructed via partial differential equation, designed by Perona and Malik [9], termed as the Perona-Malik Anisotropic Diffusion (PMAD) which is based on scale-space filtering [10]. This method gains popularity by advancing the non-linear filtering algorithm for image smoothing. Traditionally, noises are eliminated by using diffusion algorithms in use of the isotropic diffusion equation as defined as follow [11]:

$$\frac{\partial I(x, y, t)}{\partial t} = div(\nabla I) \tag{3.26}$$

Suppose the $I(x, y, t)$ denotes the input image at t stage in the continuous domain, where ∇I denotes image gradient, $(x, y, 0) : R^2 \to R^+$, (x, y) denotes the spatial coordination of the image, t denotes the time parameter. The enhanced isotropic partial diffusion equation proposed by Perona and Malik is as follows:

$$\frac{\nabla(I, x, y, t)}{\partial t} = div(g(\|\nabla I\| \nabla I)) \tag{3.27}$$

where ∇I represents gradient magnitude, $g(\nabla I)$ represents the diffusion strength functions. The diffusion function manipulates the diffusion intensity relying on the image gradient. The characteristic that is uniquely possessed by this anisotropic diffusion makes it having advantage over the conventional scale-space filtering. This characteristic is the existence of the diffusion function—the diffusion intensity varying function. This function changes accordingly depending on the image gradient. If the gradient magnitude is large, then diffusion intensity is low; whereas, if the gradient magnitude is small, the diffusion intensity is high. This complies to the final objectives of the smoothing the image: 1) the texture inside region is smoothed; 2) the object boundaries are preserved to sharpen the edges of object to preserve the details of the image. To fulfil this property of the diffusion function, Perona and Malik proposed monotonically decreasing diffusion functions as follows (2D image):

$$g_1\left(\|\nabla I\|\right) = exp\left(-\left(\frac{\nabla I(x, y, t)}{\kappa}\right)^2\right) \tag{3.28}$$

$$g_2 \left(\| \nabla I \| \right) = \cfrac{1}{1 + \left(\frac{\| \nabla I(x,y,t) \|}{\kappa} \right)^{1+\alpha}}$$

$$\text{where } \alpha > 0 \tag{3.29}$$

The 'κ' denotes an adjustable constant that is used to manipulate the 'definition of edge'. This value is conventionally decided from the noise degree in the image and the intensity of the edges in image. It is important for diffusion function to identify the edges so that diffusion strength is reduced on them with the purpose of smoothing the texture of the bone and to ease the following processes of segmentation, especially segmentation that involves clustering.

The following shows the 2D discrete implementation:

$$\frac{\partial}{\partial t} I(x, y, t) - div[g(x, y, t)^* I(x, y, t)] \tag{3.30}$$

For the relative distance, $\Delta x = \Delta y = 1$

$$\frac{\partial}{\partial t} I(x, y, t + \Delta t)$$

$$\approx I(x, y, t) + \Delta t \cdot [\Phi_{east} + \Phi_{west} + \Phi_{north} + \Phi_{south}]$$

$$+ \frac{1}{\Delta d^2} (\Phi_{northeast} + \Phi_{northwest} \tag{3.31}$$

$$+ \Phi_{southeast} + \Phi_{southwest})$$

where $\Delta d = \sqrt{2}$

3.3.2.1 Parameter-Free Diffusion Strength Function

Anisotropic diffusion is adopted in the segmentation framework with some modification so that it conforms to the desired properties P6 where the conventionally manually tuned parameter κ is made to be automated. Instead of finding κ in diffusion strength function, we used coefficient of variance. Referring to Eq. (3.27), the conventional anisotropic diffusion adopts diffusion strength function governed by image gradient and thus we have to determine the edge threshold, κ. Therefore, the aim is to modify it by using the automated noise estimation scheme from [12]. Thus $g(\| \nabla I \|)$ in Eq. (3.27) has been changed to c(q) in Eq. (3.32).

$$\frac{\nabla(I, x, y, t)}{\partial t} = div(c(q)\nabla I)) \tag{3.32}$$

$$c(q) = \frac{1}{1 + \left[q^2 - q_0^2(t) \right] / [1 + q_0^2(t)]} \tag{3.33}$$

$$q = \sqrt{\frac{var(I_{m,n})}{\mu_{I_{m,n}}^2}} \tag{3.34}$$

The $c(q)$ represents the diffusion coefficient function of SRAD where the q denotes the instantaneous coefficient of variation represented by the ratio of standard deviation as shown in Eq. (3.36), the $q_0(t)$ denotes the coefficient of variation at the time, t.

Diffusion coefficient function of speckle reducing anisotropic diffusion (SRAD), $c(q)$ is first incorporated by [13] in anisotropic diffusion SRAD to remove the multiplicative speckle noises in ultrasound image. The problem of this diffusion coefficient is the definition of standard coefficient of variance, q_0. Therefore, the decision to adopt the theory from [12] to modify $c(q)$ so that it can be automated by selecting the standard coefficient of variance, q_0 adaptively to different input image.

According to [12], the variance of noise is computed according to the expected value of local variance distribution, s^2 in the light of the variance sampling coefficient. To illustrate this idea, first looked at the relation of Chi square distribution to the definition of sampling variance distribution, s^2 at Eq. (3.35) with $N - 1$ degree of freedom [14]. Then, the definition of Chi square as a special case of Gamma distribution as shown in Eq. (3.36). After that, the inspection is shifted closer at the definition of Gamma distribution with α and β denoteing the shape parameter and amplitude parameter respectively at Eq. (3.37). The expected mean of Chi square is denoted in Eq. (3.38). Then, by computing the expected value of s^2 in Eq. (3.35), we obtained Eq. (3.39). Finally, we obtain the expected value of sample variance by substituting Eqs. (3.38) into (3.39). The conclusion is that the expected sample variance amounts to the global variance.

$$S^2 \sim \frac{(\sigma^2)^2}{N-1}(N-1) \tag{3.35}$$

$$\chi^2(N-1) \sim \gamma\left(x, \alpha = \frac{N-1}{2}, \beta = 2\right) \tag{3.36}$$

$$\gamma(x, \alpha, \beta) = \frac{1}{\Gamma(\alpha)} x^{\alpha-1} \left(\frac{1}{\beta}\right) e^{-\frac{x}{\beta}} H(x) \tag{3.37}$$

where H(x) denotes the Heaviside step function

$$E\left(\chi^2(N-1)\right) = E(\gamma(x, \alpha, \beta)) = \alpha\beta = N - 1 \tag{3.38}$$

$$E\left(S^2\right) = \frac{(\sigma^2)}{N-1} E\left(\chi^2(N-1)\right) \tag{3.39}$$

$$E\left(S^2\right) = \frac{(\sigma^2)}{N-1}(N-1) = \sigma^2 \tag{3.40}$$

This expected local variance in Eq. (3.40) shows that the global variance can be used to represent the coefficient of variance of a local sampled area. Intuitively, we can perceive that the set of pixels captured by kernel during the windowing mechanism of anisotropic diffusion represent an area with certain level of homogeneity. However, such set of homogeneities are not identical but reside in certain distribution in which they have their own variance. It has been shown in (3.40) that such variance of sample variances amounts to the global variance of the image. This fact is critical in the context of automation in anisotropic diffusion when we applied the diffusion function of SRAD as we are now able to adopt directly the variance of each input image as the q_0 into the diffusion coefficient of SRAD in Eq. (3.33). This direct adoption is critical in fulfiling criterion P2 of being high computational efficient. This get rids of the problematic manual parameter setting of value κ such as in Eqs. (3.28) and (3.29) and at the same time it fulfils the criteria of P6 and P10.

3.3.2.2 Automated Scale Selection

After automating the diffusion strength function, the central problem of anisotropic diffusion is not yet being addressed: the selection of scale. In other words, the number of anisotropic diffusion iterations before it stop diffusing the image. It can be viewed as a stopping criteria setting of anisotropic diffusion. This scale of selection is of paramount importance to the diffused image; if the stopping time is earlier than it should be, then the homogenous areas are not sufficiently smoothed; contrarily, if the stopping time is later than it should be, then the image pertinent details such as edges are smoothed out.

Conventionally, this scale is chosen manually by the user using trial and error method that is subjective, time-consuming and inconsistent and thus it violates the criteria of P7. Furthermore, it does not fulfil the P6 criterion of being automated, p10 criterion of being adaptive and P11 criterion of being optimal. Therefore, various automation schemes of selecting scale have been proposed: [15] proposed the scale selection based on energy function minimization of both computational cost and performance cost; the main weakness of this method is that the complexity in the optimization formulation. A cross-validatory statistical approach was proposed by [16]; the main weakness of this method is that the ideally-scaled image is required to perform the scheme. The use of Markov random field in scale selection has been proposed by [17]; this method suffers the main drawback of failing to represent tiny details during region segmentations in forming the super-pixel Markov random field.

All the automated scale selections found in above mentioned scale selection schemes have to compute a substantial number of extra excessive filtered images before obtaining enough information to obtain the optimum filtered image. Therefore, the crucial question towards this type of automated stopping time is that how many filtered image has to be done before it is sufficient to provide enough information to obtain the optimum filtered image. The number of filtered image needed is another parameter. The objective in establishing automated stopping time is to remove a parameter, but if the solution requires another parameter, then we would have accomplished nothing. The second problem of this type of method is that it requires huge computational time to repeats the filtered process. Therefore, we proposed a new automated stopping time (scale selection) scheme that requires only one extra filtered image and low computational complexity.

This scheme is computationally simple because it utilizes the 'by-product' from the previous procedure of establishing the diffusion strength function, i.e., the local variance. This scheme encompasses only two simple steps: (1) Build the local variance histogram (2) search the frequency of the mode of local variance that begins to deteriorate (3) stopping scale is the scale before the identified local variance in step 2. The motivation and intuitiveness of these steps are discussed below.

This hypo book of the proposed scheme is intuitive. The number of pixels that belongs to the category of relatively low local variance (potential homogenous area) belongs to the category with highest number of pixels. Therefore, when the diffusion begins to iterate, those smoothed pixels will monotonically increases until homogeneity is saturated, then the number of homogenous pixels begins to drop, and this is when the high variance pixels (edges) begins to be smoothed which is undesirable. Hence, we propose that the iteration has to be halted before the frequency of the mode of local variance begins to drop. This idea is illustrated in Fig. 3.7 by performing the diffusion on the standard image used in last chapter with all parameters preset as constant, the iteration number as manipulated

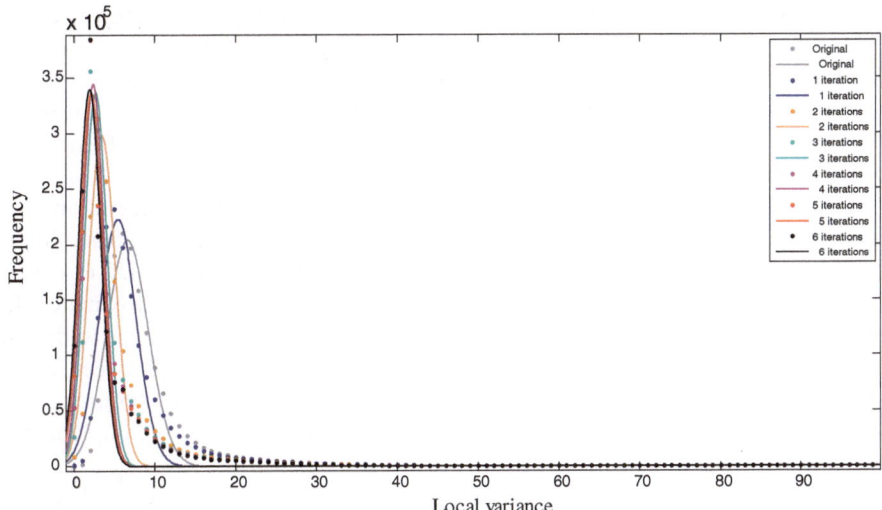

Fig. 3.7 Histogram of local variance

variable and the frequency of mode as responsive variable. The data is fitted to normal distribution to ease visual comparison and explanation.

From the figure, it is observable that the overall number of pixels in original image is shifted to the left as the iteration begins without exception. This shifting is due to the diffusion indicating that the overall regions have been smoothed and therefore the local variance is getting smaller. This smoothing process, in fact is the main purpose of diffusion. The most important observation is that the number of the distribution of local variance; from Fig. 3.7, it is observable that the mode of the local variance is monotonically increasing as the iterations begins until the fourth iterations (pink), and then it begins to drop at fifth iterations. Beyond fifth iterations, the mode begins to fluctuate instead of monotonically deceasing; this fluctuation is analogous to the concept of steady state. The reason the decision is to halt the diffusion at third iteration instead of fourth iteration in this scale selection is due to the reason that this small increment between the third and fourth is very likely attributable to the smoothing of pertinent information such as edges in the image. These observations are not randomly occurring only in certain hand bone radiographs, but occur with certainty in all hand bone radiographs in database without exception. We formulate the solution in general terms in next paragraph and explain how this scheme fulfils the desired criteria.

Let the diffused image at T-iteration to be depicted as $D(T)$. We compute the local variance of $D(T)$ and depict it as $V(T)$. Then we compute the difference between the mode of $V(T)$ and $V(T-1)$, if the difference is negative at iteration $T = m$, then the scale selection is $m-1$. This scale selection scheme conforms to the mentioned criteria. First of all, it is simple and hence has excellent performance in computational complexity and this fulfil the P2 criterion; then, it utilizes prior knowledge that edge will be smoothed out if the iterations exceed a certain number, and utilizes as well the local variance obtained in previous step of establishing the diffusion strength function; hence it fulfils the P4 criterion and in turn fulfil the P2 criterion. Besides, the scale selection is autonomous and requires no training procedure or user intervention which fulfils the P6 criterion. Lastly, it fulfils P10 criterion because it is adaptive to the image statistics instead of using any rigid thresholding technique.

Gerig et al. [18] has made an analysis on the diffusion filter integration constant, Δt, and concludes that in two dimensional discrete implementations of eight neighboring pixels, the constant range should be in between 0 and 1/7 to ensure the stability. The more Δt is to zero, the better the integration approximates the continuous case. Nevertheless, more iterations are required by the filter to diffuse the image to a certain extend. This indicates that any number in between 0 and 1/7 will not affect significantly the resultant image as long as the iterations vary accordingly. The value of the constant is set empirically as 1/7 in the implementation of the diffusion.

The flow chart of conventional anisotropic diffusion is presented in Fig. 3.8. A diffusion function has to be chosen, followed by determining a constant. This

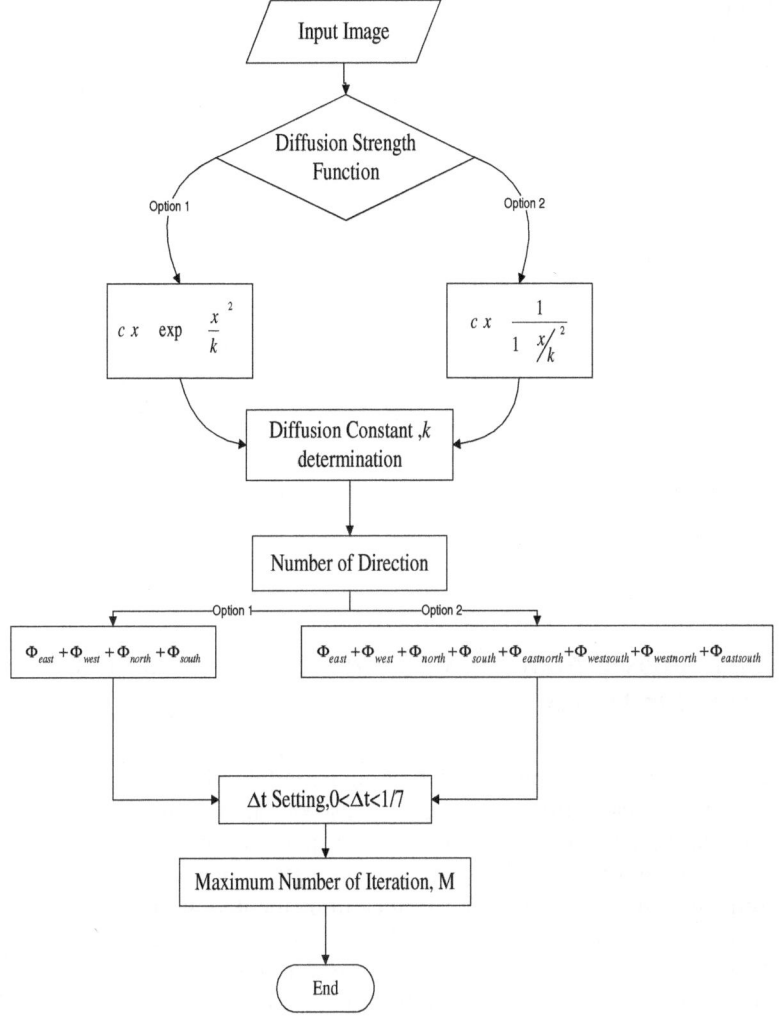

Fig. 3.8 Flow chart of conventional anisotropic diffusion

κ constant is essential to define the edges and area homogeneity; hence, different diffusion intensity is imposed on them. Subsequently, amounts of directions is to be determined: either 4 directions or 8 directions, as common practice. Then, the decision on integration constant, Δt: the effect of Δt is analogous to the approximation for continuous case using integration for discrete implementation. Finally, stopping criterion is determined: the maximum number of iteration. The flow chart of the designed automated anisotropic diffusion is presented without the need for manual parameter selection is presented in Fig. 3.9.

Fig. 3.9 The proposed
automated anisotropic
diffusion

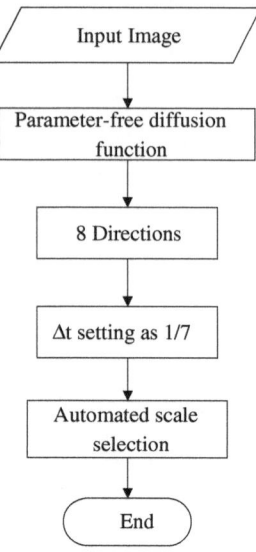

3.4 The Proposed Adaptive Crossed Reconstruction (ACR) Algorithm Design

The segmentation system begins with loading a digital radiograph of hand skeletal bone into the software. Prior to the core phase of the designed adaptive crossed recombination system, the input image is the outcome of image afer being processed by a series of pre-processing such as histogram equalization and anisotropic diffusion; the purpose of the former implementation is to standardize the radiograph so that radiographs from different machine of x-ray will be standardized; the purpose of the latter implementation is to enhance the histogram graph by diffusing the intensity value within the bone while preserving the sharp edge of the bone.

After the pre-processing, the image will enter the main phase of segmentation. The algorithm begins with division accordingly to quadruple division algorithm in Sect. 3.4.2. The purpose of dividing the image into sub-image is to fulfil adaptive property; the image contrast is not uniform and the degree of changes is increased throughout the whole image from the phalange until the carpal. Dividing the image into sub-image is able to resolve this problem by providing a suitable environment for subsequent step of k-mean clustering. Henceforth, the divided sub-image is termed as block.

The k-mean clustering is performed on each block. With the k-means unsupervised clustering technique, the x-ray image is clustered into two and three groups which represent bone, soft-tissue region and bone, soft-tissue region and background with 'k' value equals to two and three respectively. The implementation of this clustering process is carried out by optimizing an objective function, in this case, to minimize a squared error function, as shown in (3.41).

$$\sum_{j=1}^{K}\sum_{i=1}^{n}||x_i^{(j)} - c_j||^2 \tag{3.41}$$

where n is the number of data, K is the clusters number, $||x_i^{(j)} - c_j||^2$ is the distance measurement between the data point $x_i^{(j)}$ representing the intensity value of image pixels and the cluster center c_j computed from calculating the mean of each group of pixels. It indicates the distance difference from the data point to the cluster centre. First, the histogram of the input image is computed. Two clusters centre will be chosen randomly among the data points followed by constructing the distance difference between each data point and the cluster centre. Subsequently, each data point will be assigned to one of the two clusters centre depends on the distance difference. When all the data points are assigned to cluster centre, the new cluster centre is then recalculated by computing the mean of each cluster group. These steps iterate until the objective function achieve a minimum value.

3.4.1 Clustering Algorithm Applied in the Proposed Segmentation Framework

Algorithm

(A) **Input:**

Data set (image pixels intensity) $= \{x_i\}_1^n$, where n represents the total number of pixels in the image.

(B) **Initiation:**

(i)   Set the clusters number 'k', $1 \leq k \leq 3$ where k is an integer.
(ii) Initiate the cluster center, $C_j(T), 1 < j \leq k, T = 0$
(iii) Set the tolerant error value $'\sigma'$

where T = number of k-means iterations, Initiation value: k = 1, T = 0, j = 1, and i = 1

(C) **Computation of Euclidean Distance**

The Euclidean Distance between intensity value and cluster center value is computed as follow:

$$D_{ji}^{(T)} = \sum_{j=1}^{K}\sum_{i=1}^{n}w_{ji}\left\|x_i^{(j)} - C_j^{(T-1)}\right\|$$

$$\tag{3.42}$$

(D) **The Assignment of Pixel's Intensity Value To Cluster Center**

$$w_{ji} = \begin{cases} 1, argmin_{j=1}^{k} D_{ij}^{(T)} \\ 0, Otherwise \end{cases} \tag{3.43}$$

(E) **Cluster Center Recalculation**

$$C_j^T = \frac{\sum_i^n w_{ji}^T x_i}{\sum_i^n w_{ji}^T} \tag{3.44}$$

(F) **Stopping Criteria**

Iterate until $E(T)$ is less than tolerant error value '∂'

$$E(T) = \left\| C_j^T - C_j^{T-1} \right\| \leq \partial \tag{3.45}$$

(G) **Iterations**

Repeats with $T \leftarrow T+1$ in step 2 if stopping criteria is not achieved

(H) **Cluster Center Replacement**Fill the pixels belong to cluster with lower value of cluster center with zero intensity and remain the pixels intensity that belong to cluster with higher value of cluster center.

$$x_i^{(j)} = \begin{cases} 0, argmin_{j=1}^{2} Cj \\ x_i, otherwise \end{cases} \tag{3.46}$$

(I) **Cluster Center Replacement**
   (i) Repeat the process with $k = 2$
   (ii) Repeat the process with $k = 3$

(J) **Cluster Center Replacement**

Fill the pixels belong to clusters with lower value of cluster center with zero intensity and remain the pixels intensity that belongs to cluster with highest value of cluster center.

$$x_i^{(j)} = \begin{cases} x_i, argmax_{j=1}^{3} Cj \\ 0, otherwise \\ 0, otherwise \end{cases} \tag{3.47}$$

Grey Level Co-occurrence Matrix (GLCM) proposed by Haralick et al. [19] suggested a way to describe texture of image by statistical measurement. A GLCM is a matrix that contains the probability of gray scale i occurs together with the gray level j at a specific distance, in a specific direction. The distance and direction are named offset at times and different value of offset will give different indication for the texture of image. Gray scale levels used in this book is divided into four in order to increase the algorithm computing speed:

$$I(x,y) = \begin{cases} 1, if\ 0 \leq I(x,y) \leq \dfrac{argmax_{xy}^{columnrow}I(x,y)}{4} \\ 2, if\ \dfrac{argmax_{xy}^{columnrow}I(x,y)}{4} \leq I(x,y) \leq \dfrac{argmax_{xy}^{columnrow}I(x,y)}{2} \\ 3, \dfrac{argmax_{xy}^{columnrow}I(x,y)}{2} \leq I(x,y) \leq \dfrac{argmax_{xy}^{columnrow}I(x,y)}{\frac{4}{3}} \\ 4, if\ \dfrac{argmax_{xy}^{columnrow}I(x,y)}{\frac{4}{\pi}} \leq I(x,y) \leq argmax_{xy}^{columnrow}I(. \end{cases} \tag{3.48}$$

where $I(x,y)$ represents pixels gray level intensity at coordinate(x, y), x represents column and y represents row. The second order statistical measure that relevant in this book is shown in Eqs. (3.49) and (3.50).

$$Contrast = \sum_{i,j=0}^{N-1} P_{i,j}(i-j)^2 \tag{3.49}$$

$$Homogeneity = HOM = \sum_{i,j=0}^{N-1} \frac{P_{i,j}}{1+(i-j)^2} \tag{3.50}$$

where $P_{i,j}$ represents the probability a group of spatial related pixel intensity occur in the image. Among the twelve statistical equations proposed by Haralicks [19], homogeneity is a suitable one for the purpose in the segmentation framework in order to choose a uniform region of image. Homogeneity, also called 'Inverse Difference Moment' is an inversion to the contrast. While computing the contrast, the weight of element increases when distance of element from diagonal of the GLCM increases. Inversely, the weight of element decreases as the distance of elements from diagonal increases. In short, the weight of contrast is $(i-j)^2$, on the other hand, the weight of Homogeneity is $\frac{1}{1+(i-j)^2}$.

The texture analysis in the proposed ACR k-means clustering method will be conducted twice. First the texture analysis implemented on selecting the number of k in k-means clustering, both images for k equals to two and k equals to three will be computed, however only one of them will be used in reconstruction step, the criteria is based on the homogeneity value. The concept is that between every set of two resultant segmented images in each block, the one with higher homogeneity will be selected.

$$B(i) = B(i,k)$$
$$where\ k = argmax_{k \in 23}(HOM(B(i,k))) \tag{3.51}$$

where $B(i)$ represents the i-th quadruple divided block of the image, $HOM(B(i,k))$ represents homogeneity value in $B(i,k)$. All the blocks are then combined using

Eq. (3.52) to form the final segmented image. The maximum number of block depends on the quadruple division algorithm in Sect. 3.4.2.

$$Final\ Segmented\ Image = \bigcup_{i=1}^{max\ i} B(i, k) \qquad (3.52)$$

Next we have to optimize the result by instilling some rules and intelligence for the machine incorporating with homogeneity to adaptively opt for the suitable size of block in every region using the proposed automated block division scheme as explained in following sections.

3.4.2 Automated Block Division Scheme in Adaptive Segmentation

In designing the adaptive clustering algorithm, two decisions are crucial: Firstly, the size of the block and secondly the robustness of the designed clustering algorithm under different circumstances. The common desired properties of choosing the blocks' size and designing clustering algorithm are automated (fulfilment of P6) and adaptive (fulfilment of P10). Being automated implies that there is no involvement of user-specified parameters or any manual manipulation to avoid the drawback of subjectivity and inconsistency of making decision by users. Being adaptive indicates that each decision is made according to its given context on the ground that the quantities and characteristic involved are dynamic and should not be treated similarly using single preset parameter value so that it is adaptive to different environment or context.

The conventional unsupervised clustering fulfils the requirement of being automated, but not the second requirement which is robust under different conditions. The inconsistent robustness of the conventional clustering algorithm is attributable to various factors: the number and 'location' of initial centroids, relatively unequal proportional size, densities of the natural structure of data, the shape of the natural structure of data, the disturbance of outliers.

In the context of hand bone x-ray radiograph, the main problem is the unpredictable proportional number of pixels in each of the underlying natural partitions of background, soft-tissue region, cancellous bone and cortical bone while performing the adaptive segmentation. For example, the number of pixel belong to cortical bone region could not be predicted in some block where most of the region in the corresponding block are background and soft-tissue region. In this kind of block, the conventional unsupervised clustering might fail to optimize the objective function and lead to the undesired visual effect after partitioning the pixels inside the block into four clusters. Therefore, there is a need to analyze the proportional number of pixels in each of the four partitions before processing them using unsupervised clustering. In other words, extra information about the block is required to have optimized performance in the unsupervised clustering of each block because the context of each block is different. This information can be acquired through a design of a series of continual decision steps in the context of hand bone. The information obtained from the analysis is used to determine whether the size of the block is suitable and therefore should remain or it is unsuitable and therefore should be altered. More details on blocks' size are discussed later.

One major drawback of adopting adaptive block (sub-image) segmentation is the choice of the block. The effectiveness of the adaptive segmentation decreases as block's size increases due to less influence on non-uniform illumination. However, if the size is smaller than it should be, then the content of the block might not contain adequate information for accurate segmentation and furthermore, it is more prone to noise influence. Therefore, choosing a suitable block's size is the main challenge of adopting adaptive segmentation, not to mention choosing the suitable size automatically. To the best of our knowledge, there is not yet any analytically proven solution to tackle this problem.

We suggested that the size of adaptive division block should depend on the performance of clustering algorithm used in separating the bone image into compact bone, spongy bone, soft-tissue and background by relying on the result of cluster properties to ensure the resultant pixel intensity clusters from unsupervised clustering technique are meaningful and useful in subsequent procedures of segmentation. Meaningful clusters refer to group of resultant clusters that are able to reveal the natural structure of the data.

In the context of hand bone image processing, the structures of the data represent the anatomical structure's region of the hand bone. This process of analyzing the properties of the resultant cluster to determine the next step of segmentation procedure can be perceived as an intelligence of the algorithm in assessing the current condition and then making appropriate decision to optimize the result (fulfilment of P8). If the resultant clusters are considered meaningless after analysis, then this result indicates that the current size of the adaptive division block is not suitable and in turn it indicates that there is better option in terms of the size of the block and hence appropriate adjustment is needed. On the other hand, if the resultant clusters are considered meaningful after the analysis, then this result indicates that the current size is acceptable and it is permitted to proceed to next procedure of segmentation. We propose that a similar division process is performed in those meaningless blocks and this process repeats until all current blocks have achieved suitable size. Next we discuss the reason why 'suitable' is used to describe the size rather than 'optimized'.

The reason of not adjusting the size until it optimizes the objective function of cluster validity is that being too accurate is computationally infeasible and thus it would affect its practical usage especially in real-time computational environment when dealing with high definition radiographs or when it is implemented in any usage where temporal factor is an issue; for example, the implementation in hand bone segmentation in computer-aided skeletal age scoring system. One of contributions in the proposed block division scheme is that even relatively high accuracy of clustering can be achieved without sacrificing the computational feasibility by designing a feedback procedure of anatomical structure outline assessment process. In other words, there is trade-off between the accuracy and the computational costs. Therefore, in the designated system of segmentation, we aim at achieving the relatively high performance on both accuracy and the computational cost which could not be achieved inherently by any conventional unsupervised clustering.

3.4.2.1 The Framework of the Proposed Scheme

The main advantage of this proposed division algorithm in comparison with the conventional adaptive segmentation is its ability to divide and stop dividing itself automatically without user-specified parameters. Instead, it depends only on the proposed steps of decision that is specially designed in the context of hand bone radiograph to reveal the content. The second advantage is its ability to divide irregularly adaptively to the content within the blocks automatically instead of the rigid division in any local thresholding scheme. These two abilities greatly improve the performance of adaptive segmentation. The drawbacks of conventional methods and the importance of these two abilities are discussed below.

Conventional adaptive segmentation requires user to manipulate parameters such as the number of divisions of input image into regions with the intent of adaptively processing different regions with different thresholds. The number of divisions depends on the resolution and the total pixels' number of the input image. Users have to implement the trial-and-errors methods in order to find the suitable number of divisions. This implementation is prone to human errors, time-consuming and laborious; thus, it is not practical in hand bone segmentation where automation is a must and time taken has to be short enough to be practically implemented.

Besides having to set the parameter, the conventional adaptive segmentation always divides the input images into regular blocks' size to simplify the process. This simplification decreases the adaptability of the adaptive segmentation. The reason lies in the fact that region in each block has different distribution range of pixel intensity and different proportional number of pixels. Regular identical blocks size leads to artifacts as only some of the blocks produce desired effect and the rest are not. Contrarily, irregular blocks size enables better tuning of parameters in different regions in input image and thus it increases the adaptability and produce better result. The adaptability is crucial to analyze the region locally rather than analyzing it globally in terms of image context. The analysis depends on the content of the block and the resultant content of the block after being processed by unsupervised clustering based on a series of designed analysis which we will discuss later.

Firstly, the input image is divided into four blocks (quadrisection). Secondly, the proposed cluster validity is implemented on each block. Depending on the result of analysis, the block will either continue the quadrisection or remains as it is. This process is simple but the benefits it obtains are influential in terms of segmentation effectiveness. In fact, the simplicity is the key advantage itself. The dividing mechanism is analogous to the 'separate and merge' procedure of region growing excluding the merging part.

The mechanism can be described more effectively by using diagram and flow-chart as below. To ease the explanation, nomenclatures used are defined. The top left corner's block is named as TL-1.0, the top right corner's block is named as TR-2.0, the bottom left corner's block is named as BL-3.0, the bottom right corner's block is named as BL-4.0. If the TR-1.0 is quadrisected into four equal blocks, then the top left corner's block of TR-1.0 is named as TR-1.1; the top right corner's block of TR-1.0 is named as TR-1.2; the bottom left corner's block of TR-1.0 is named as TR-1.3 and the logic goes on. For example, if the BR-4.0 is quad-risected, then the bottom right corner's block of it is named as BR-4.4, and if again BR-4.4 is further quadrisected, then for instance, the bottom left corner's block of it is named as BR-4.4.3 and so on.

The Fig. 3.10a is the input image. Figure 3.10b shows the quadrisected input image. In Fig. 3.10c, TL-1.0 remains according to the result of cluster validity;

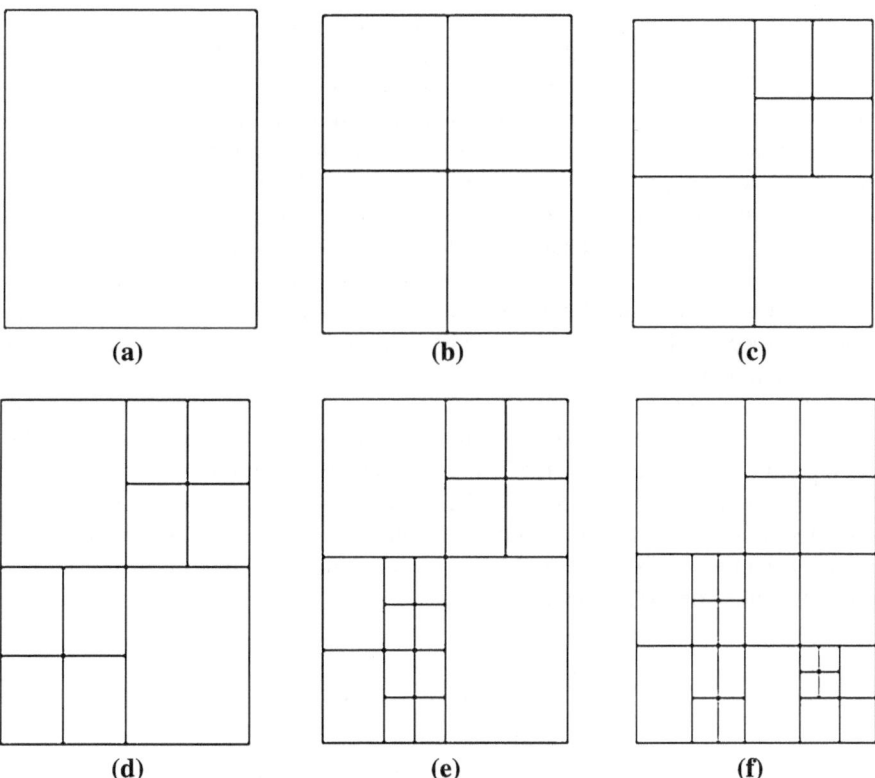

Fig. 3.10 The illustration of quadruple Block's division algorithm using an arbitrary example

cluster validity begins on TR-2.0 and decides that TR-2.0 needs to be further quadrisected to achieve better result. In Fig. 3.10d, each of the TR-2.1, TR-2.2, TR-2.3, TR-2.4 remains according to the result of cluster validity and therefore proceed to BL-3.0. In Fig. 3.10e, BL-3.1 and BL-3.3 remain whereas BL-3.2 and BL-3.4 are quadrisected according to the cluster validity; all sub-blocks of BL-3.2 and BL- 3.4 which are BL-3.2.1, BL-3.2.2, BL-3.2.3, BL-3.2.4, BL-3.4.1, BL-3.4.2, BL-3.4.3, BL-3.4.4 remain as decided in accordance to cluster validity on each of them and therefore proceed to BR-4.0. In Fig. 3.10f, BR-4.1, BR-4.2, BR-4.3 remain whereas BR-4.4 is quadrisected according to the result of cluster validity; each sub-blocks of BR-4.4 undergoes cluster validity and result in BR-4.4.2, BR-4.4.3, BR-4.4.4 remain as they are whereas BR-4.4.1 have to be further quadrisected. Finally, all of the sub-blocks of BR-4.4.1 remain and hence complete the process of quadruple block's division algorithm.

3.4.2.2 The Mechanism of the Automated Fuzzy Quadruple Division Scheme

After performing the ACR clustering on each block, a series of analysis were then performed to determine whether to further dividing the block or remain as it is. The central concept of the proposed method is to provide best 'environment' for the previous ACR algorithm to optimize the result. The decision is based on a large amount of a priori knowledge: (1) the nature of the pixel intensity distribution, (2) number of pixel in each partition, (3) the location in histogram. Four simple metrics were defined to describe the nature of pixels. These metrics were designed so that they made use of the 'by-product' of previous unsupervised clustering in the proposed ACR algorithm to give some clues in order to predict the pattern in the blocks. All these clues are crucial to play the role. Human a priori knowledge is inserted into the algorithm of automated block division to govern the dividing process. This a priori knowledge determines when to keep dividing and when to stop dividing.

Firstly, we defined the Normalized Difference Expectation, NDE between two clusters, as the difference between the expectation pixel intensity of the clusters C_1 and C_2 from Sect. 3.4.1 when $k = 2$. The expectation value is defined by Eq. (3.53). Let $c_{ij} \in C_j$, where $j = 1,2$, $i = 1$ to total members number of C_j, denoted as $max(n(C_j))$. This metric as follows is to measure the distance relation between two clusters, where M(I) represents the maximum grey level in input block, I, in this context, the M(I) is 256:

$$E\left(C_j\right) = \frac{1}{max(n(C_j))} \sum_{i=1}^{max(n(C_j))} C_{ij} \qquad (3.53)$$

$$NDE\left(C_1, C_2\right) = \frac{\|E\left(C_1\right) - E\left(C_2\right)\|}{M(I)} \qquad (3.54)$$

Why this distance matters in the context of ACR algorithm? It is because there are many possible contents being captured within the region in the divided block and NDE provided us some clues about the contents of the block:

1. Case 1: The block might be fully filled up with radiograph background dark pixels or almost fully filled up with radiograph background dark pixels and a small region of non-background dark pixels.
2. Case 2: The block might be fully filled up with radiograph non-background relatively brighter pixels or almost fully filled up with radiograph non-background pixels and a small region of background dark pixels.
3. Case 3 is that the block might be almost or equally filled up with relatively brighter non-background pixels and dark background pixels.

To determine which of the above cases is most likely to be true, another important metric that we have to define is the Normalized Cluster Location, *NCL*, of both clusters; the idea is very intuitive; if both nearby clusters tend to be situated on the bright intensity region in histogram, then the case 2 is more likely to be true; if both nearby clusters tend to be situated on the darker intensity region in histogram, then the case 1 is more likely to be true. If the mean brightness of both clusters is moderate, then case 3 is the most likely case. We attempted to predict both cluster overall location using the following metric:

$$NCL\,(T) = \frac{\left[E\,(C_1) + E\,(C_2)\right]/2}{M\,(I)} \tag{3.55}$$

With *NDE* defined in Eq. (3.54) and *NCL* defined in Eq. (3.55), we can now estimate the content of the corresponding block by classifying case 1 from case 2. The reason we desired to separate case 1 from case 2 is that block that reside in case 1 category requires no dividing anymore, whereas blocks that belong to case 2 category need to undergo further analysis to determine whether to halt or to continue the dividing. We analyze case 2 by computing the Homogeneity Discrepancy Test (HDT) of segmented regions from ACR algorithm for k = 2 and k = 3; if the discrepancy level is low, then the dividing process of the block has to continue and vice versa. This analysis is computationally simple (fulfil P2), and making use of homogeneity score from Eq. (3.50) in previous ACR segmentation process (fulfil P4).

$$HDT\,(i,k) = \frac{\|HOM\,(B\,(i,k=2) - HOM\,(B(i,k=3)\|}{argmin_{k \in 23} HOM\,(B\,(i,k)} \tag{3.56}$$

If the *HDT* show low discrepancy, then we have to continue the quadruple dividing process because the strength of evidence is not strong enough to terminate the dividing process.

Now we have to identify the case 3. The rationale is that if the block is not from case 1 and case 2, then it belongs to case 3. However, within the set of blocks that belong to case 3, there are many variations as there are numerous ratios of bright area and dark area within a block. To evaluate the ratios, the most intuitive approach is to compute the fraction between bright and dark area give that the block neither belongs to case 1 nor belongs to case 2. In other words, once we gather enough evidence to determine both clusters are far apart, then the ratios of the number belong to both clusters give us another clue about how well will the ACR algorithm segment the content within the block; if the clue suggests that the ACR algorithm could not segment the content with confidence, then the quadruple block dividing scheme continues.

Let $n(C_j)$ denotes total number of pixel reside in j-th clusters; in this specific context, $j = 1, j = 2$. Inspired by Shannon Entropy, the Clustering Efficacy Ratio CER $(n(C_1), n(C_2))$ is defined as (3.57), where the range of the interval ratio $\frac{n(C_1)}{n(C_1)+n(C_2)}$ is inclusive of unity but exclusive of zero, denoted as $\frac{n(C_1)}{n(C_1)+n(C_2)} \in (0 1]$. Total of $\frac{n(C_2)}{n(C_1)+n(C_2)}$ and $\frac{n(C_1)}{n(C_1)+n(C_2)}$ amounts to unity, hence $\frac{n(C_2)}{n(C_1)+n(C_2)}$ can be represented as $1 - \frac{n(C_1)}{n(C_1)+n(C_2)}$; as a result (3.57) is defined in terms of $\frac{n(C_1)}{n(C_1)+n(C_2)}$ as represented by (3.58).

$$
CER\,(n(C_1), n(C_2)) = -\,[\frac{n\,(C_1)}{n(C_1)+n(C_2)}log_2\frac{n(C_1)}{n(C_1)+n(C_2)} \\
+ \frac{n\,(C_2)}{n(C_1)+n(C_2)}log_2\frac{n(C_2)}{n(C_1)+n(C_2)}]
\tag{3.57}
$$

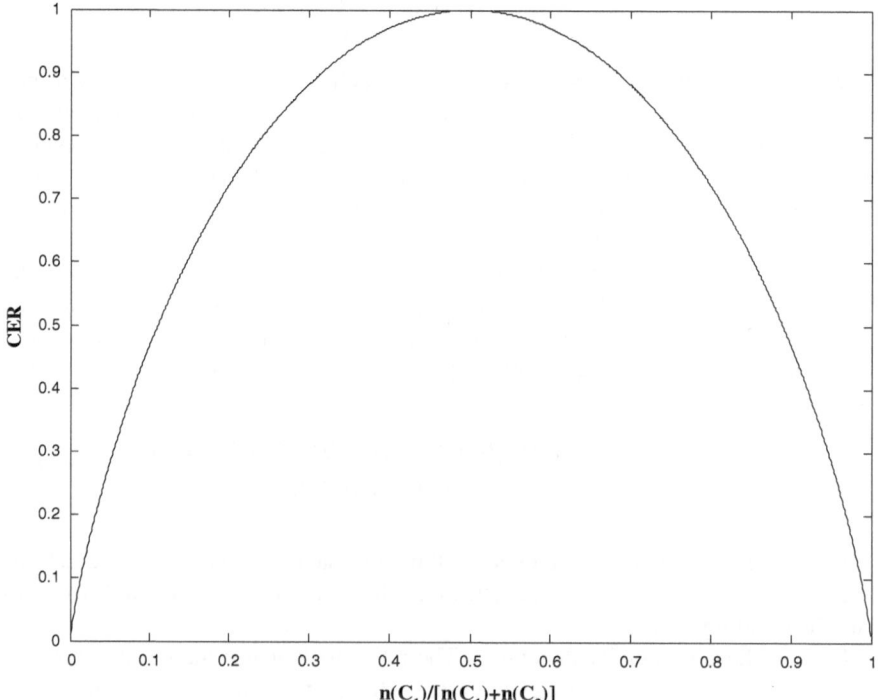

Fig. 3.11 The *CER* function relative to cluster ratio

$$CER\left(n\left(C_1\right), n\left(C_2\right)\right) = -\left[\frac{n\left(C_1\right)}{n(C_1)+n(C_2)}log_2\frac{n(C_1)}{n(C_1)+n(C_2)}\right]$$
$$+\left[1-\frac{n\left(C_1\right)}{n(C_1)+n(C_2)}\right]log_2\left[1-\frac{n(C_1)}{n(C_1)+n(C_2)}\right]$$

$$(3.58)$$

CER map the ratio $\frac{n(C_1)}{n(C_1)+n(C_2)}$ onto the interval (0 1] to model the strenght of the equality ratio of both clusters: If the $n(C_1)$ amounts to $n(C_2)$, then a value of unity is associated with it; on the contrary, as the $n(C_1)$ significantly deviates from $n(C_2)$, then a value of approaching zero is associated with it, however, it will never become zero as the minimum number of pixels reside in any cluster is one pixel. This enables the logarithm function to be utilized in this modeling. The CER function versus ratio $\frac{n(C_1)}{n(C_1)+n(C_2)}$ is shown in Fig. 3.11.

The CER function provided a numerical evaluation of the block content about the ratio between the first cluster and the second cluster. If the ratio is relatively high or relatively low, the CER results in low value, then the dividing process is

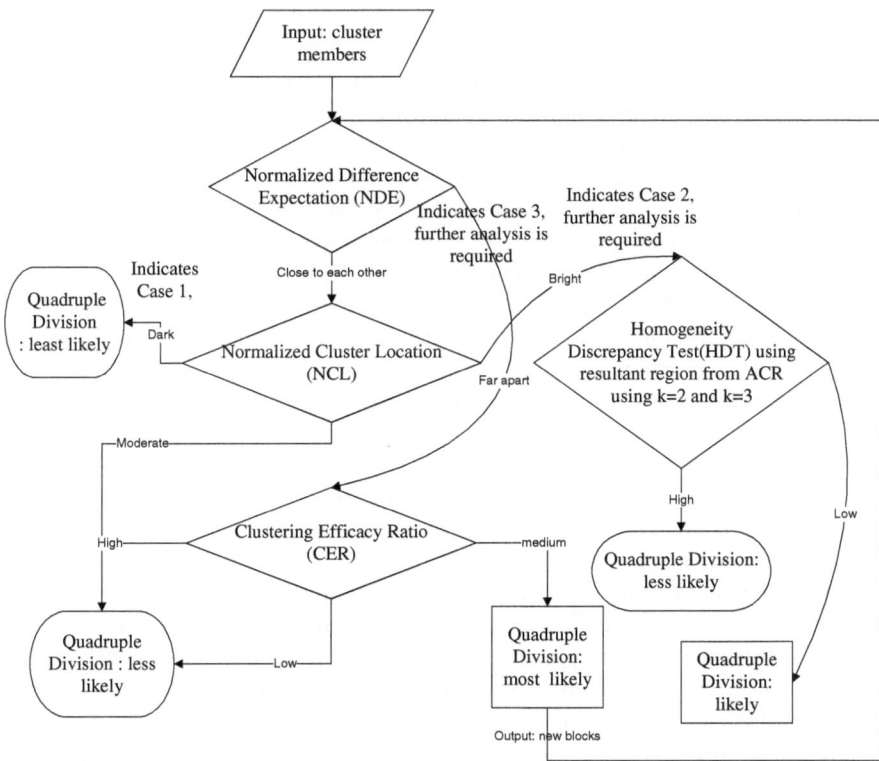

Fig. 3.12 The decision tree for quadruple division scheme

terminated, whereas, if the ratio is in the moderate range, then the *CER* results in high value, and the dividing process continues.

The procedures of mentioned process can be viewed as constructing a decision tree for classifying blocks that should continue undergoing quadruple dividing and blocks that should stop undergoing quadruple dividing process. This kind of decision tree outperforms other sophisticated cluster validity in this context in terms of computational demand. Beside, this kind of decision flow generates comprehensive rules for users to understand and manipulate. The main component in this decision flow has been defined. The entire process can be better illustrated via a decision flowchart shown in Fig. 3.12.

As we can observe in Fig. 3.12, there are many vague and ambiguous adjectives such as 'close to', 'far apart', 'high', 'low', 'dark', 'bright' 'moderate', 'medium', 'less likely', 'most likely' and 'least likely'. These words are common terms in daily linguistic communication to describe intuitive a priori knowledge. In terms of semantics, these words carry no precise definition and are very vague to be measured in exact numerical number. First of all, the question is that why we describe the relations of the cluster in vague words instead of exact numerical numbers, for example, if the *NCL*, below 0.3 or any numerical threshold, then it is considered as 'near 0'? Basically, there are three reasons:

1. To fulfil the important criterion of P8 in which we desire to impose human visual perception intuition into this automated scheme so that it behaves autonomously while still manage to resemble human decision making process and cognitive ability; human do not use numerical numbers, instead, human utilize intuition and a priori knowledge to perceive and interpret an image, and yet we human beings are capable of classifying extremely complicated objects from a very chaotic image. In the case of *NCL*, it is not logical for human to interpret clusters that produce *NCL* below 0.3 are considered as 'near 0', and *NCL* value such as 0.31 is considered as not close to 0. The cognitive process of human simply do not behave this way and thus any pre-set thresholding is not effective in imitating human cognitive visual ability. Moreover, pre-set thresholds, as mentioned in previous chapter, is not adaptive to the block's content and thus does not fulfil criterion P10 in handling variability.

2. The cause and effect are investigated through empirical study and human observation; the analytic function and precise numerical relation between the cause and effect in the complex system containing pixel intensity clusters, optimum block size, resultant segmented image, are not yet available. When these analytic function and precise definition do not exist, only vague linguistic semantics can model the human observation in the complex system.

3. The exact numerical solution or threshold is not necessarily required to perform crucial decision in the complex system. Decision making using approximation gain significant computational efficacy advantage that is paramount in fulfiling criterion P2 which is a principal advantage of the proposed scheme compared to other computationally demanding scheme such as active shape model and active appearance model. Therefore, if approximations involve, vague descriptive terms are inevitable.

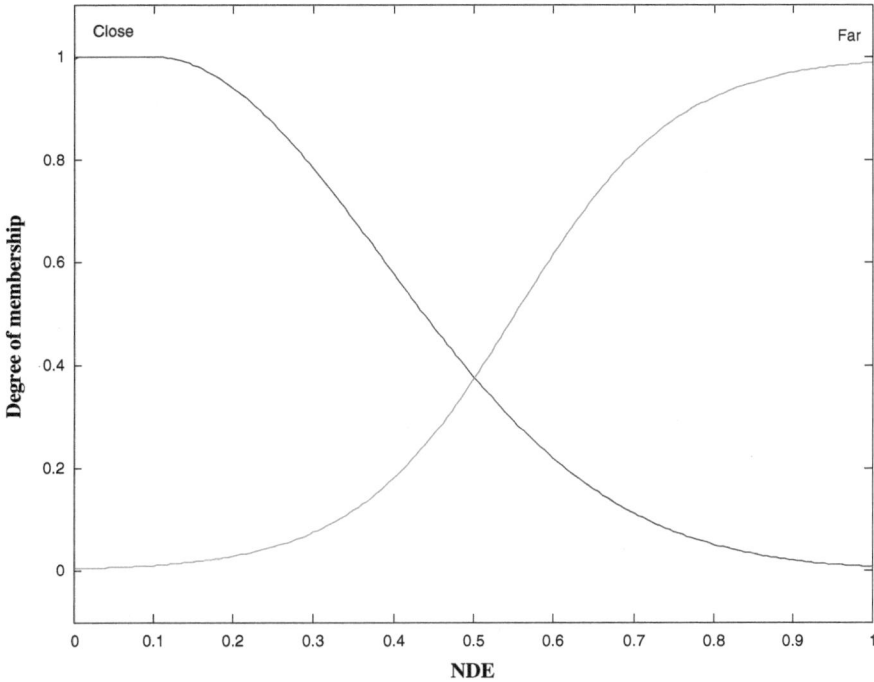

Fig. 3.13 Membership function of *NDE*

The above reasons explained why vague words were used instead of precise numerical value. Now we explain how to process the vague definition. To evaluate the vague quantities to come up with a decision is not a trivial challenge. This process is designed in fuzzy system framework, which is a well-known logic framework in processing these interconnected multistage vague decisions in Fig. 3.12 in order to produce a desired response to establish the automated control on the quadruple division scheme.

The formulation of *NDE*, *NCL*, HDT, and *CER* defined in Eqs. (3.54), (3.55), (3.56) and (3.57), respectively, can be perceived as inputs in establishing fuzzy membership functions in fuzzy logic controller using intuitive approach. Each fuzzy membership of Eqs. (3.54), (3.55), (3.56) and (3.57) is illustrated graphically as follow in Figs. 3.13, 3.14, 3.15 and 3.16, respectively to elucidate the relation of the input and the designed membership strength. Figure 3.17 shows the output quadruple division's membership function.

The decision rules described in Fig. 3.12 can be perceived as linguistic rules in fuzzy logic controller. The fuzzy inference mechanism system adopted is the most commonly used Mamdani and Assilian [20] implication rules:

1. Represent the 'and' antecedent connective using min (intersection) operator; represent the 'or' antecedent connective using max (union) operator. Let *A* and *B* denote two fuzzy sets on the X universe, and x denotes any member

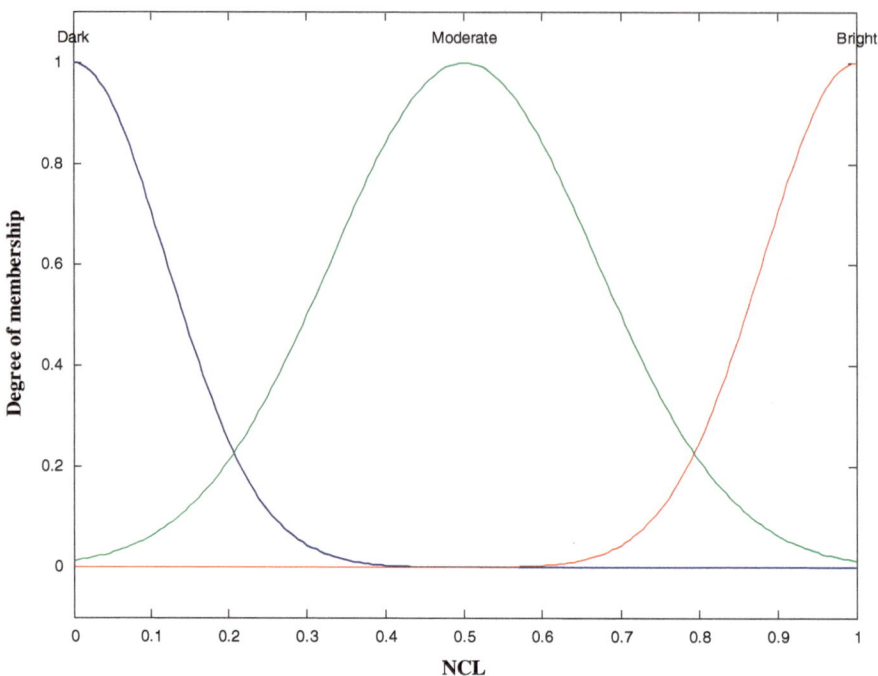

Fig. 3.14 Membership function of *NCL*

reside in the X universe. The fuzzy relations of intersection and union in Mandani implication rules are represented as follow:

$$\mu_{A \cup B}(x) = \max[\mu_A(x), \mu_B(x)] \tag{3.59}$$

$$\mu_{A \cap B}(x) = \min[\mu_A(x), \mu_B(x)] \tag{3.60}$$

2. In antecedent part of fuzzy linguistic rules, evaluate the minimum or product of all membership functions as the resultant antecedent membership function.
3. Evaluate the consequent membership function and the resultant antecedent membership function from 2) using the either Mamdani max–min inference method (clipped method) or max-product inference method (scaled method).

The overview of the fuzzy system framework can be illustrated using Fig. 3.18. The seven linguistic inference rules presented in Fig. 3.12 are restated in linguistic semantic rules as follows:

1) Rule 1

Rule antecedent	If	*NDE*	is close
	and	*NCL*	is dark
Rule Consequent	then	Quadruple division	is least likely

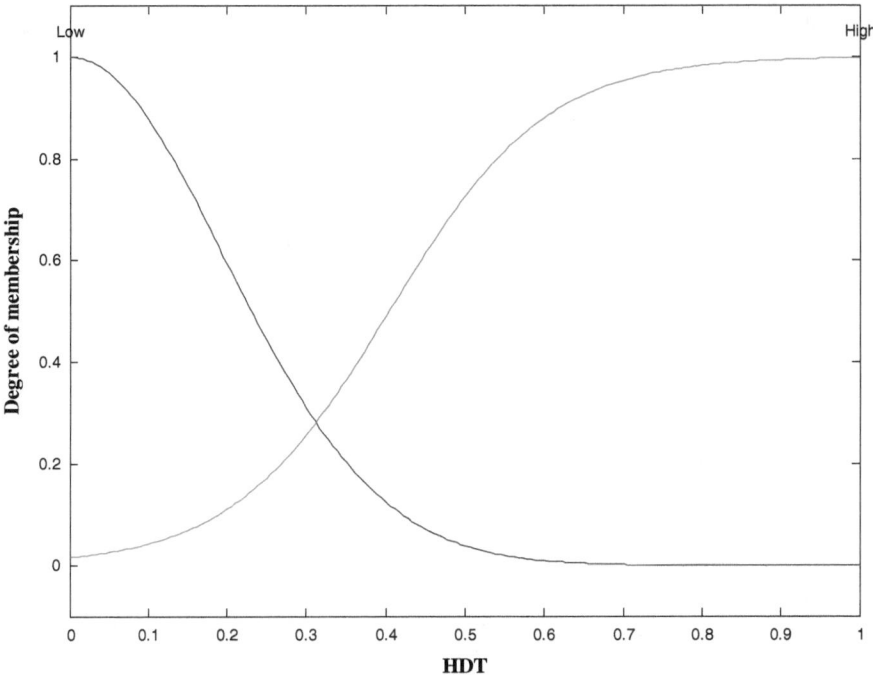

Fig. 3.15   Membership function of *HDT*

2) Rule 2

Rule antecedent	If	*NDE*	is close
	and	*NCL*	is moderate
Rule consequent	then	Quadruple division	is less likely

3) Rule 3

Rule antecedent	If	*NDE*	is close
	and	*NCL*	is bright
	and	*HDT*	is high
Rule consequent	then	Quadruple division	is less likely

4) Rule 4

Rule antecedent	If	*NDE*	is close
	and	*NCL*	is bright
	and	*HDT*	is low
Rule consequent	then	Quadruple division	is likely

5) Rule 5

Rule antecedent	If	*NDE*	is far
	and	*CER*	is high
Rule consequent	then	Quadruple division	less likely

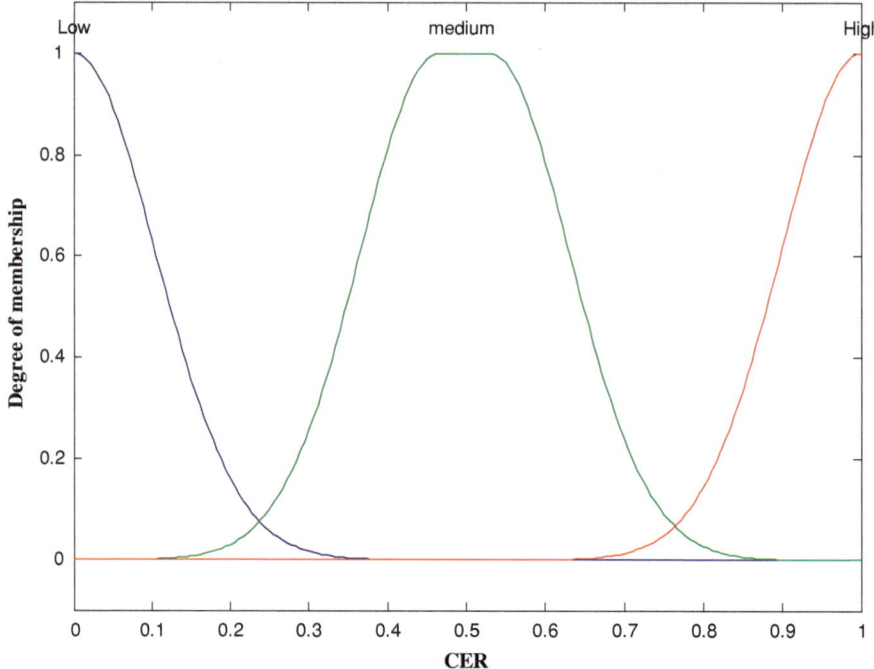

Fig. 3.16   Membership function of *CER*

6) Rule 6

Rule antecedent	If	*NDE*	is far
	and	*CER*	is low
Rule consequent	then	Quadruple division	Less likely

7) Rule 7

Rule antecedent	If	*NDE*	is far
	and	*CER*	is medium
Rule consequent	then	Quadruple division	most likely

For every rule, after evaluation, will produce rule output. The process of combining of all membership function contributed by each rules output into a single fuzzy set is called the 'aggregation' process. This process conclude all the rules output. In Mandani method, this process is accomplished by using the max operator.

The way human observe, understand, analyze and model is vague and uncertain. However, when it comes to decision making, crisp logic is more reasonable to human. The decisions human make are mostly in binary nature. For example, in the context of quadruple division, the final decision is binary: Terminate or continue the division process. In other words, when human considered the input

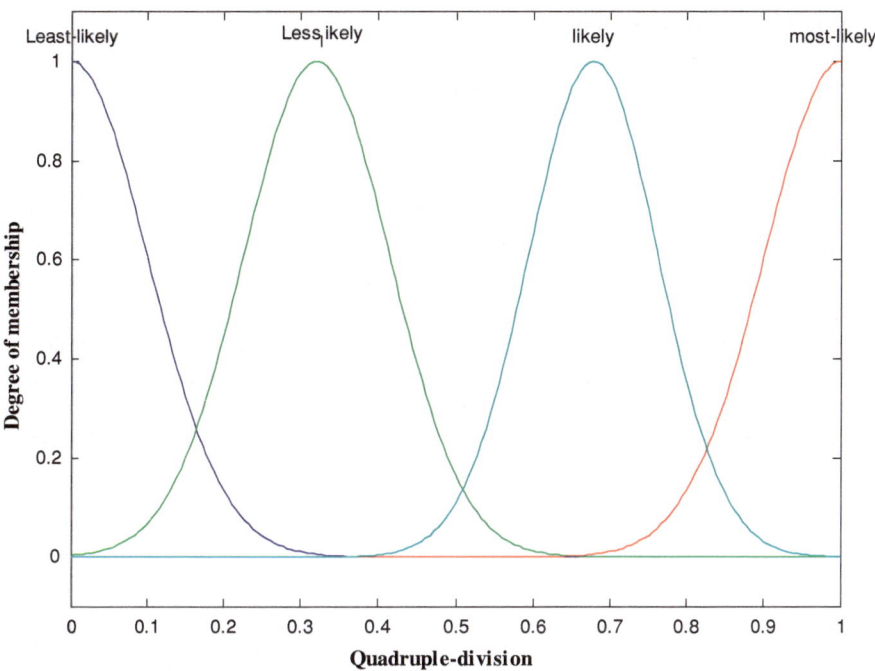

Fig. 3.17   Quadruple division output membership function

Fig. 3.18   Overview of the fuzzy inference system

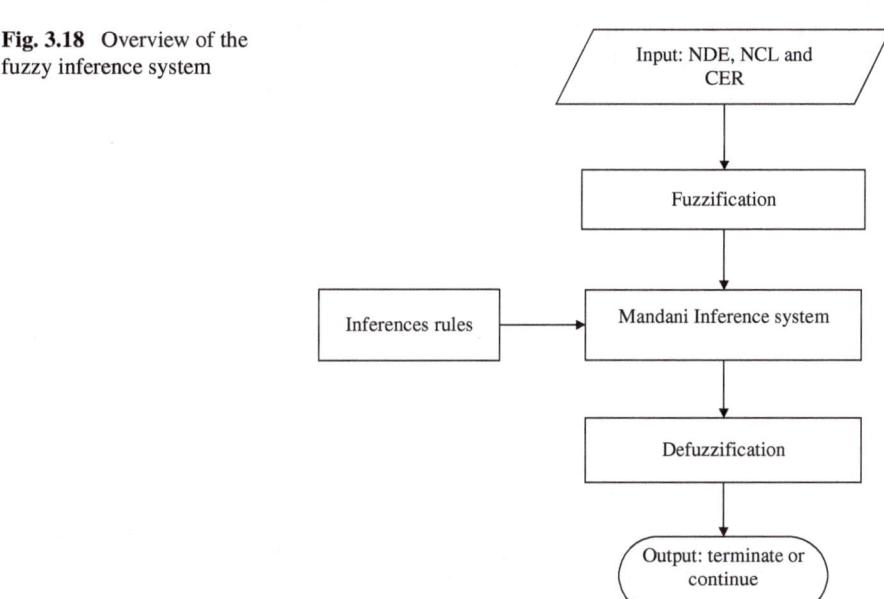

variables of any observation, fuzzy concept is useful to describe the vague nature of reality and to comprehend the natural linguistic semantic of human; however, when we want to take a subsequent action, crisp logic is useful to make a binary decision in accordance to the interpreted fuzzy information. This resembles the decision making process of a human expert.

This process of transforming fuzzy output set into crisp output set is termed as 'defuzzification'. The concept is similar to rounding up from decimal numerical number into integer numerical number in mathematics. Analogously, in the context of fuzziness, this process can be viewed geometrically as rounding up the range value of the output membership function to the nearest vertex of the corresponding membership function. There are five common fuzzifiers: smallest of max, largest of max, centroid of area, bisector of area, and mean of max. In the proposed application, centroid of area method is adopted to find the centroid of the aggregated output membership function, $c*$, defined mathematically as follows:

$$c^* = \frac{\int \mu_A(c) \,.cdc}{\int \mu_A(c)\, dc} \tag{3.61}$$

Decide to continue if

(Quadruple division $|NDE, NCL, HDT, CER)$

$> 1 - (Quadruple\ division|NDE, NCL, HDT, CER)$ (3.62)

otherwise decide to terminate

For better understanding, the idea is illustrated with an arbitrary numerical example. Let the $NDE = 0.2$, $NCL = 0.4$, $HDT = 0.6$, $CER = 0.8$. The resultant computed centroid is shown in Fig. 3.19 which amounts to 0.292. In this particular case, according to classification rule, the logical decision is to be terminated as in the decision rule below. This is the quadruple division is controlled automatically using fuzzy logic by incorporating human expert knowledge (Fig. 3.20).

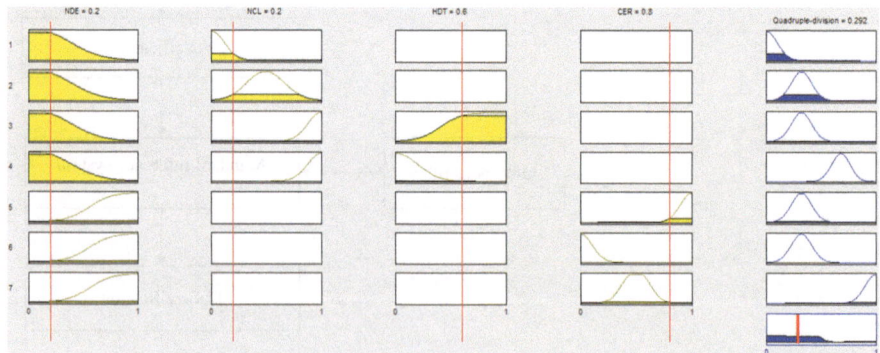

Fig. 3.19 Overall input–output relationship using max–min composition and centroid defuzzifier. The red line in the aggregated output in bottom right denotes the centroid position of the aggregated output membership area

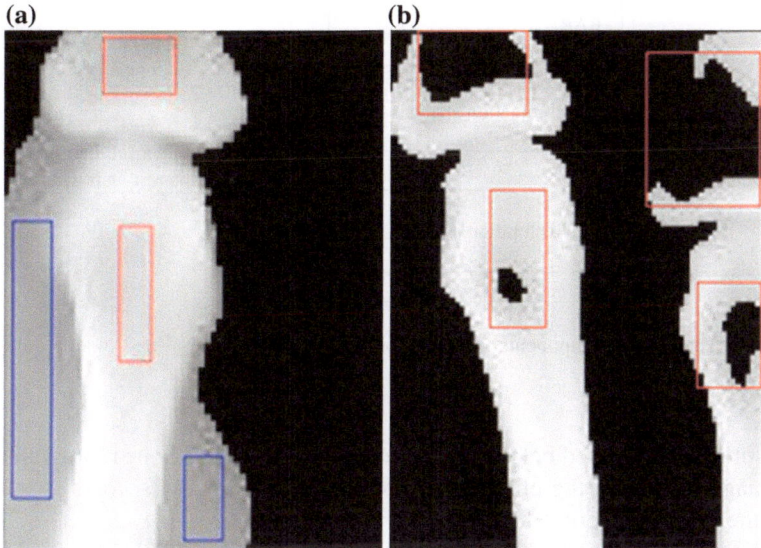

Fig. 3.20 Two arbitrary blocks in segmented bone image. (**a**) This bone image contains extra regions that should be eliminated. The range of pixels intensity within the red rectangular boxes (spongy bone pixels) is overlapping with range of pixels intensity within the blue rectangular boxes (soft-tissue pixels). The former group of pixel needs to be maintained, and the latter group of pixels need to be eliminated (**b**) this bone image has been over-segmented and the regions that undergo this artifacts are regions in spongy bone as expected, which are pixels within the red rectangle. Those missing pixels have to be restored

3.5   Quality Assurance Process

Despite previous schemes, not all pixels are perfectly classified via ACR k-mean clustering in quadruple division framework is correct due to the inherent nature of hand bone radiograph that the pixel intensity range of soft-tissue region and the spongy bone is overlapping. Therefore, most of the errors are due to incorrect classification of soft-tissue region and spongy bone. For instance, the pixels that belong to soft-tissue region have been wrongly classified as spongy bone and hence they have been included in the segmented hand bone image; the pixels that belong to the spongy bone have been wrongly classified as pixels in soft-tissue region and hence have been excluded in the segmented hand bone image. The extra pixels in the former case need to be eliminated and the wrongly excluded pixels in the latter case need to be restored to obtain an accurate segmented hand bone image. It is not trivial at all to detect the wrongly classified pixels and perform the restoration and elimination. Therefore, we proposed a restoration and elimination system by using the Bounded-area Restoration and Non-bounded Area Elimination (BARNAE) with some carefully chosen and designed features. The detail of the features chosen and the illustration of the Restoration and elimination

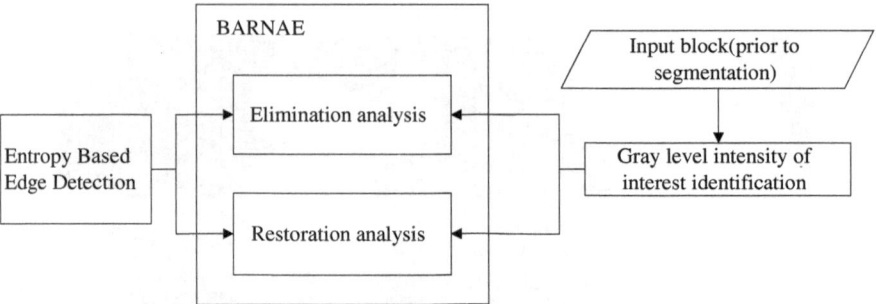

Fig. 3.21 The main components in the bounded-area restoration and non-bounded area elimination

procedures are discussed below using two instances of hand bone radiographs. The first image illustrated the elimination procedure whereas the second image illustrated the restoration procedure.

The entire restoration and elimination scheme can be clearly illustrated using the flowchart in Fig. 3.21 which shows the scheme is using input of detected edges on the pixels that is of identified gray level intensity of interest. The intensity of interest identification is discussed in Sect. 3.5.1; the entropy based edge detection is discussed in Sect. 3.5.2; the BARNAE algorithm is discussed in Sect. 3.5.3.

3.5.1 Gray Level Intensity of Interest Identification for Elimination

The purpose to identify the range of pixel intensity is to reduce the number of pixels that requires further analysis, only pixels that are in high risk of wrongly classified are proceed to analysis to serve two goals to determine whether to remain the pixel or eliminate it as segmented pixel (elimination in this context means setting the corresponding pixel to intensity 0).

Firstly, the histogram of the segmented bone pixels is plotted and interpolated using curve fitting technique by a sum of multiple Gaussian models:

$$g(x) = \sum_{i=1}^{n=3} A_i e^{\frac{-(x-x_i)^2}{2e_i^2}} \tag{3.63}$$

The parameter n is set as 3 as there are 3 groups of pixels which comprise of a group of compact bone pixels, a group of background pixels, and a group of pixels which belong to either soft-tissue region or spongy bone region. Next, to find the third group of pixels, the second and third inflection points are obtained and the pixels within these two limits are classified as the targeted group of pixels which would be analyzed to perform the restoration and elimination.

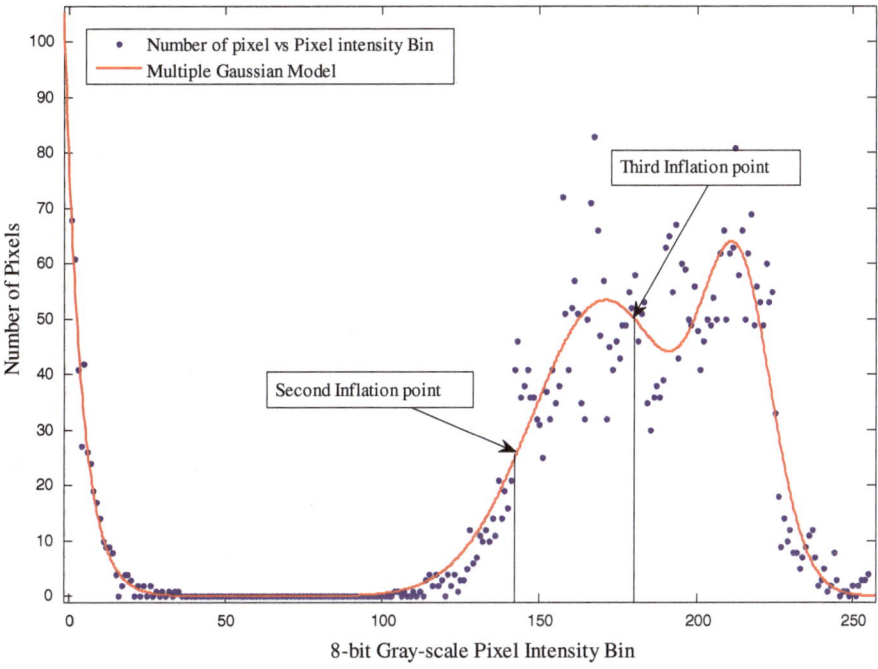

Fig. 3.22 The histogram of segmented bone image of Fig. 3.20a and its interpolated Gaussian graph with second inflation point and third inflation point

For instance, the inflection points of the interpolated Gaussian graph of the arbitrarily selected segmented bone image of Fig. 3.20a, are 144 and 182. Thus, those pixels belong to this newly defined group, which are pixels with pixel intensity between 144 and 182, are going to be analyzed, whether it is to be eliminated or remained. Pixel intensity value more than 182 is considered as compact bone group and pixel intensity value more than 144 is considered as background group. The histogram, interpolated Gaussian model and the inflection points are shown in Fig. 3.22.

For segmented bone image of Fig. 3.20b, the inflection points of the interpolated Gaussian graph are 197 and 242. Thus, those pixels belong to this newly defined group, which are pixels with pixel intensity between 197 and 242, are going to be analyzed, whether to be eliminated or restored. The histogram, interpolated Gaussian model and the inflection points are shown in Fig. 3.23.

3.5.2 Hand Bone Edge Detection Technique Using Entropy

Owing to the nature of hand bone image, conventional image processing techniques are unable to detect the edges easily (this statement will be justified in next chapter). Therefore, we propose a technique where the by-product of anisotropic

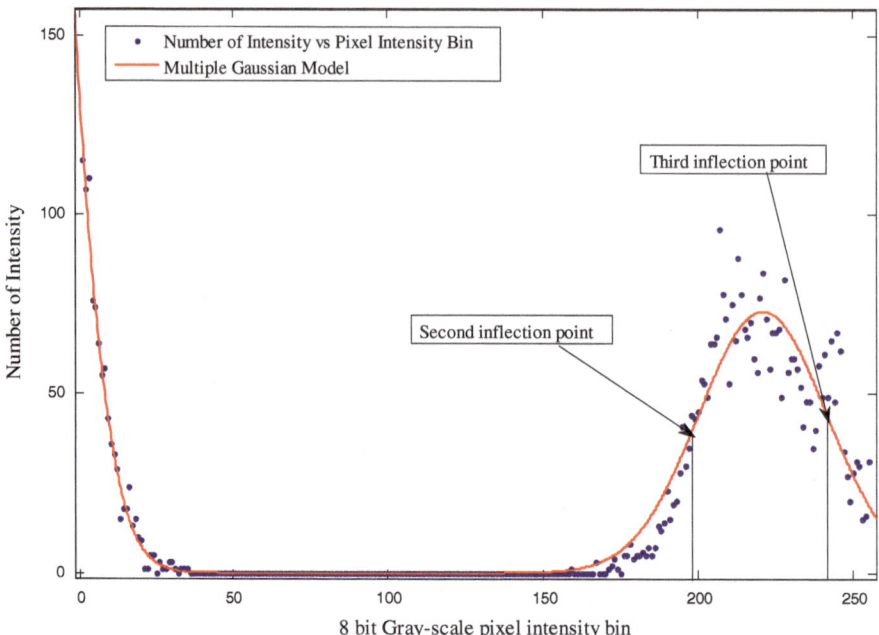

Fig. 3.23 The histogram of segmented bone image of Fig. 3.20b and its interpolated Gaussian graph with second inflation point and third inflation point

diffusion is utilized as the pre-processing of the edge detection, follows by GLCM entropy based edge detection technique, then lastly, utilize the unsupervised clustering technique to separate the noises and edge pixels. This proposed technique share a similar concept to ACR segmentation and therefore it could utilize the by-products of previous processing techniques to prevent the wastage of information and increase the computational efficiency in detecting the edges. Besides, through comparisons with other conventional method, this technique has been proven to be more robust, the comparison result is shown in next chapter. The algorithm details of the proposed edge detection method are described next:

An intuitive solution is to use the entropy as a quantitative measurement of information content, originated from information theory of Shannon [21]. This measurement enables the assessment of the randomness of the pixel intensity distribution in hand bone image. Mathematically, it is represented by the following equation:

$$Entropy = -\sum_{i}^{N-1}\sum_{j}^{N-1} P_{(i,j)} ln P_{(i,j)} \qquad (3.64)$$

where $P_{(i,j)}$ depicts the probability a group of spatial related pixel intensity occur in the image. N depicts the adopted gray level. The entropy reaches its maximum value, which is 0.5 only when all $P_{(i,j)}$ are equally distributed. On the contrary, the entropy reaches its minimum value, which is 0, only when all $P_{(i,j)}$ are equal to zero. This indicates that the more occurrences of different group spatial related pixel intensity, the larger the entropy. Whereas, if there is less occurrences of different group spatial related pixel intensity, the smaller the entropy. This entropy concept of measuring randomness is analogous to measuring the local variance. Therefore, instead of computing the entropy value, the local variance value obtained in previous process to increase the computational efficiency is adopted. The only difference is that the local variance in Eq. (3.40) has been normalized.

$$VI\,(x,y) = D_{m-1}(x,y) \tag{3.65}$$

The main problem that impedes this scheme of straight-forward edge detection is the nature of hand bone image contains high variations on bone texture which will affect this removal of pixel. Therefore, the anisotropic diffused image from pre-processing is utilized rather than the input image. Let the $(m\text{-}1)$-iterated diffused image as $D_{m-1}(x,y)$, each single pixel in the diffused image undergo an local variance transformation using windowing technique, and let the transformed image as $VI(x,y)$. This produced an image where the pixels inside the bone have been transformed to low intensity value as shown in Fig. 3.24.

After transforming to normalized local variance images, there are noises and artifacts. To separate the spurious edges from the actual edges, k-mean clustering on edge strength is adopted. As a result, the detected edges pixels for Fig. 3.20a, b after being denoised was shown in Fig. 3.25a, b respectively.

In order to illustrate the relative position of edges and the segmented hand bone for better understanding, the detected edges were combined with segmented hand bone image. Figure 3.26 showed the result of the combination where Fig. 3.20a is combined with Figs. 3.25a, 3.20b is combined with Fig. 3.25b to produce Fig. 3.26a, b respectively.

Fig. 3.24 Normalized local variance transformation from diffused bone images of Fig. 3.20a, b

(a)   **(b)**

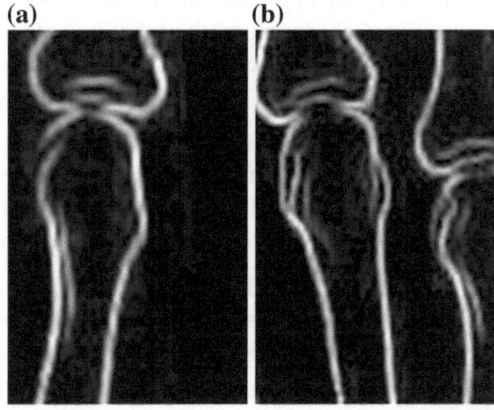

Fig. 3.25 The final
detected hand bone edges of
Fig. 3.20a, b after denoising
using k-mean clustering

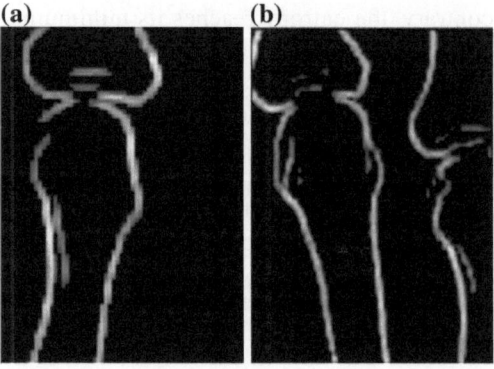

Fig. 3.26 The Combination
of hand bone and detected
edges

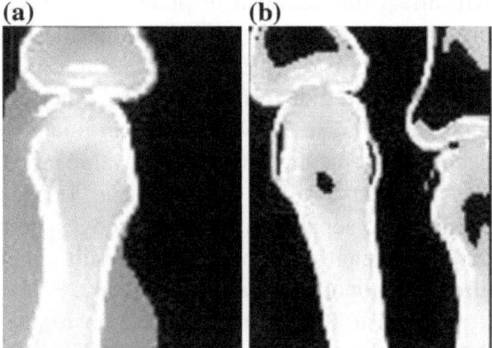

3.5.3 The Area Restoration and Elimination Analysis

Similar to any existing segmentation frameworks, the proposed framework is by
no means a perfect segmentation framework because it is the inherent weakness of
conventional clustering method that it contains no pixel spatial information. These
imperfections or artifacts after segmentation are always due to two reasons:

1. False segmentation of spongy bone and therefore a restoration process was set up to
   identify unfilled bounded areas (lost data) and fill it with original bone pixel value.
2. False labeling of soft-tissue regions as bone and therefore a elimination process
   was set up to identify these over-segmentation artifacts followed by eliminating
   it and replacing it back to soft-tissue region.

The steps in the BARNAE: Step in part (A) demonstrates the labeling process in
each direction. Step in part (B) explains the stopping criteria. Step in part (C) defines
the recognition of bounded area, for it a noise or lost data. The entire process mentioned above is repeated in step in part (D). Last step involves the filling in the lost
data or elimination of noise. The overview of the BARNAE is illustrated in Fig. 3.27.

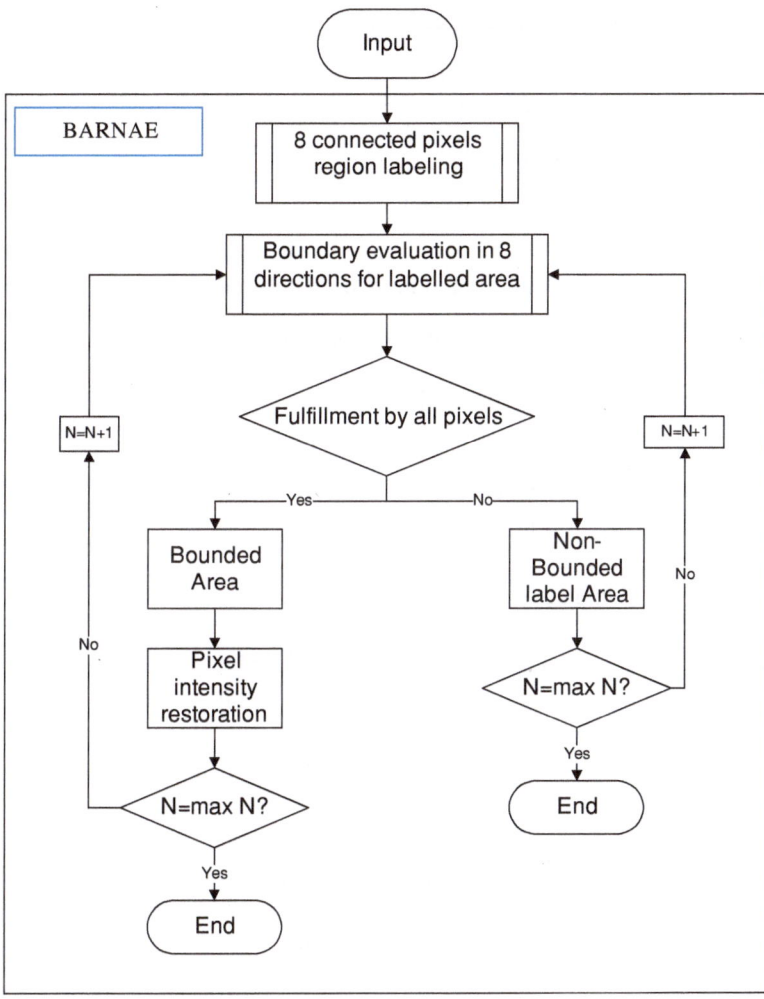

Fig. 3.27   The overview of BARNAE algorithm

The boundary evaluation for labeled area algorithm in Fig. 3.27 is described below from Eqs. (3.66) to (3.76).

Boundary evaluation process:

(A) Pixel Evaluation

      (i)   Evaluate each pixel in n-label for $0°$

$$f(x,y) = \begin{cases} 0, I^b_{x+k,y} & \text{if } b = n, \\ 1, I^b_{x+k,y} & \text{if } b \neq n, \end{cases}$$

where   (3.66)

$$k = 1 \rightarrow L$$

L = maximum number of column $-x$

(ii)   Evaluate each pixel for 45°

$$f(x,y) = \begin{cases} 0, I^b_{x+m,y-k} & \text{if } b = n, \\ 1, I^b_{x+m,y-k} & \text{if } b \neq n, \end{cases}$$

where   (3.67)

$k = 1 \rightarrow y \rightarrow 1$

$m = 1 \rightarrow L$

L = maximum number of column −x

(iii)   Evaluate each pixel in n-label for 90°

$$f(x,y) = \begin{cases} 0, I^b_{x,y-k} & \text{if } b = n, \\ 1, I^b_{x,y-k} & \text{if } b \neq n, \end{cases}$$

where   (3.68)

$k = 1 \rightarrow y - 1$

(iv)   Evaluate each pixel in n-label for 135°

$$f(x,y) = \begin{cases} 0, I^b_{x-m,y-k} & \text{if } b = n, \\ 1, I^b_{x-m,y-k} & \text{if } b \neq n, \end{cases}$$

where   (3.69)

$k = 1 \rightarrow y - 1$

$m = 1 \rightarrow x - 1$

(v)   Evaluate each pixel in n-label for 180°

$$f(x,y) = \begin{cases} 0, I^b_{x-m,y} & \text{if } b = n, \\ 1, I^b_{x-m,y} & \text{if } b \neq n, \end{cases}$$

where   (3.70)

$k = 1 \rightarrow y - 1$

$m = 1 \rightarrow x - 1$

(vi)   Evaluate each pixel in n-label for 225°

$$f(x,y) = \begin{cases} 0, I^b_{x-m,y+k} & \text{if } b = n, \\ 1, I^b_{x-m,y+k} & \text{if } b \neq n, \end{cases}$$

where   (3.71)

$k = 1 \rightarrow L$

$m = 1 \rightarrow x - 1$

L = maximum number of row −x

(vii)   Evaluate each pixel in n-label for 270°

$$f(x,y) = \begin{cases} 0, I^b_{x,y+k} & \text{if } b = n, \\ 1, I^b_{x,y+k} & \text{if } b \neq n, \end{cases}$$

where

$k = 1 \to L$

L = maximum number of row −x

$\qquad\qquad\qquad\qquad\qquad\qquad\qquad\qquad\qquad\qquad$ (3.72)

(viii)   Evaluate each pixel in n-label for 315°

$$f(x,y) = \begin{cases} 0, I^b_{x+m,y+k} & \text{if } b = n, \\ 1, I^b_{x+m,y+k} & \text{if } b \neq n, \end{cases}$$

where

$k = 1 \to L1$

$m = 1 \to L2$

L1 = maximum number of row − y

L2 = maximum number of column − x

$\qquad\qquad\qquad\qquad\qquad\qquad\qquad\qquad\qquad\qquad$ (3.73)

(B) Stopping criteria:

$$\left[f(x,y) = 1 \right] \cup n = N$$

where N = Maximum Label

$\qquad\qquad$ (3.74)

(C) Verification of bounded area for n-label

$n -$ labelled area

$$= \begin{cases} \text{bounded area, } if\ \frac{\sum^{\max_x}_x \sum^{\max_y}_y f(x,y) I^n_{x,y}}{\text{total number of pixels in n−label}} = 1 \\ \text{non - bounded area, otherwise} \end{cases}$$
$\qquad\qquad$ (3.75)

(D) Repeat the process with n = n+1, where n denotes the label number of pixels in image.

(E) Fill the pixels belong to bounded area with original value/background value of pixel intensity of the corresponding coordinate in the image:

$$I_{x,y} = I^n_{x,y}$$
$\qquad\qquad$ (3.76))

The BARNAE algorithm analyzes each of the third group pixel found using multiple Gaussian model. The result of the analysis determines whether the pixel is wrongly classified. If the pixel should belong to soft-tissue region, but has been wrongly classified as spongy bone region, then this pixel will be eliminated; in the contrary, if the pixel should belong to spongy bone region, but has been wrongly classified as soft-tissue region, then the pixel will be restored. The analysis is based on extending eight directions of the pixel until it 'touches' the edge pixels, if all direction of the pixel 'touches' the edges, then the pixel is considered as a bounded pixel within the bone and hence it is restored. In contrast, if at least one direction of the pixel does not 'touch' the edge pixel after extending, then the pixel is considered as a non-bounded pixel within the bone and hence it is eliminated. The resultant image of Fig. 3.20a, b after BARNAE are shown in Fig. 3.28a, b, respectively.

3.6 Summary

In previous chapter, the weaknesses of conventional segmentation methods have been identified. This concludes the desired segmentation criteria in order to guide the mechanism of the proposed framework of segmentation. The segmentation is performed to partition the hand bone from its background and soft-tissue region in the beginning of this chapter. The challenges of hand bone segmentation is the overlapping intensity between the soft-tissue region and the spongy bone region within the hand bone. A segmentation framework consisting of three main modules has been proposed and implemented to solve the problem: pre-processing, ACR in fuzzy quadruple division framework and quality assurance process. Each of them plays equally important role in tackling the challenge. Pre-processing consists of two main components: histogram equalization and anisotropic diffusion. The proposed histogram equalization, MBOBHE is specially customized to protrude the features of the hand bone and to curb the problem of uneven illumination of the radiograph to prepare a radiograph invariant of illumination for the subsequent processings of ACR segmentation. Besides, the anisotropic diffusion

Fig. 3.28 The resultant blocks after BARNAE

(a)   **(b)**

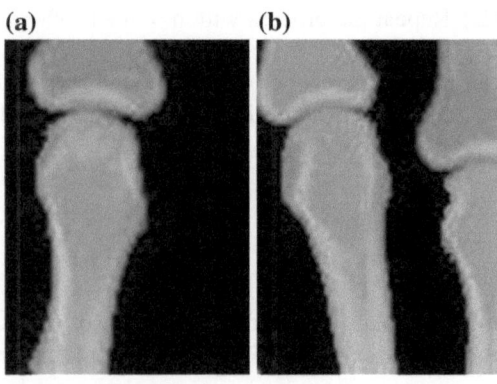

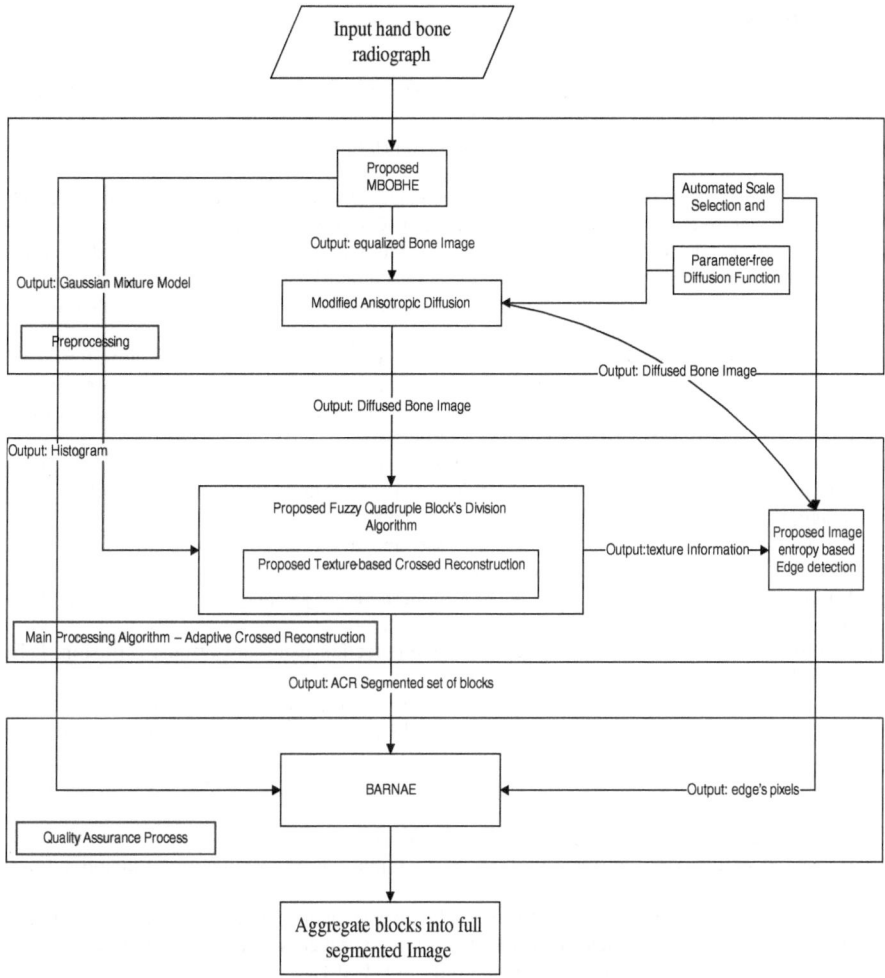

Fig. 3.29 The overview of the proposed hand bone segmentation framework

has been designed to be automated in terms of the scale selection and the diffusion strength function; this is important to assure that the entire framework has least human intervention. The diffused and equalized radiograph will then enter the stage of ACR segmentation using the fuzzy quadruple block's division scheme. After that, each block undergoes quality assurance process to restore lost image data and eliminate abundant image data. Finally, all the blocks are aggregated to form the final segmented hand bone radiograph that are suitable to be used as input radiograph for any computer-aided skeletal age scoring system. Figure 3.29 summarizes all the abovementioned modules to provide an overview of the entire segmentation framework.

References

1. Nakib A, Oulhadj H, Siarry P (2010) Image thresholding based on Pareto multiobjective optimization. Eng Appl Artif Intell 23:313–320
2. Chen S-D, Rahman Ramli A (2004) Preserving brightness in histogram equalization based contrast enhancement techniques. Digital Signal Process 14:413–428
3. Yu W, Qian C, Baeomin Z (1999) Image enhancement based on equal area dualistic sub-image histogram equalization method. IEEE Trans Consum Electron 45:68–75
4. Cao F, Huang HK, Pietka E, Gilsanz V (2000) Digital hand atlas and web-based bone age assessment: system design and implementation. Comput Med Imaging Graph 24:297–307
5. Sim KS, Tso CP, Tan YY (2007) Recursive sub-image histogram equalization applied to gray scale images. Pattern Recogn Lett 28:1209–1221
6. Kim M, Min C (2008) Recursively separated and weighted histogram equalization for brightness preservation and contrast enhancement. IEEE Trans Consum Electron 54:1389–1397
7. Kim T, Paik J (2008) Adaptive contrast enhancement using gain-controllable clipped histogram equalization. IEEE Trans Consum Electron 54:1803–1810
8. Seungjoon Y, Jae Hwan O, Yungfun P (2003) Contrast enhancement using histogram equalization with bin underflow and bin overflow. In: Proceedings of international conference on image processing, 2003 (ICIP 2003) (ed), vol 1, pp I-881-4
9. Perona P (1989) Anisotropic diffusion processes in early vision. In: Multidimensional signal processing workshop, 1989. Sixth edition, p 68
10. Witkin A (1983) Scale-space filtering. In 8th international joint conference of artificial intelligence. (ed), Vol 2, pp 1019–22
11. Perona P, Malik J (1990) Scale-space and edge detection using anisotropic diffusion. IEEE Trans Pattern Anal Mach Intell 12:629–639
12. Aja-Fernández S, Vegas-Sánchez-Ferrero G, Martín-Fernández M, Alberola-López C (2009) Automatic noise estimation in images using local statistics. Additive and multiplicative cases. Image Vis Comput 27:756–770
13. Yongjian Y, Acton ST (2002) Speckle reducing anisotropic diffusion. IEEE Trans Image Process 11:1260–1270
14. Fuller W (2009) Sampling Statistics. Vol. 560, Wiley, London
15. Capuzzo Dolcetta I, Ferretti R (2001) Optimal stopping time formulation of adaptive image filtering. Appl Math Optim 43:245–258
16. Papandreou G, Maragos P (2005) A cross-validatory statistical approach to scale selection for image denoising by nonlinear diffusion. In: IEEE computer society conference on computer vision and pattern recognition (CVPR'05) (ed), Vol. 1, IEEE computer society, pp 625–30
17. Sun J, Xu Z (2010) Scale selection for anisotropic diffusion filter by Markov random field model. Pattern Recogn 43:2630–2645
18. Gerig G, Kubler O, Kikinis R, Jolesz FA (1992) Nonlinear anisotropic filtering of MRI data. IEEE Trans Med Imaging 11:221–232
19. Haralick RM, Shanmugam K, Dinstein IH (1973) Textural Features for Image Classification. IEEE Trans Syst, Man Cybern 3:610–621
20. Mamdani EH, Assilian S (1975) An experiment in linguistic synthesis with a fuzzy logic controller. Int J Man Mach Stud 7:1–13
21. Shannon CE (1948) A mathematical theory of communication. Bell Syst Tech J 27:379–423

Chapter 4
Result and Discussion

Abstract In this chapter, the functionality of proposed MBOBHE and the modified anisotropic diffusion will be justified. Both techniques create a suitable environment for the subsequent segmentation modules by reducing the disturbance of uneven illumination, sharpening the edge and smoothing the texture; at the same time, MBOBHE is capable of enhancing the features of ossification sites to improve the performance of computerized BAA. After that, an analytical comparison of AAM segmentation framework with the proposed segmentation framework is performed.

4.1 Introduction

As mentioned earlier, the radiograph used in all the experiments in this chapter were adopted from hand bone online database http://ipilab.org which comprises both genders in four populations which are Caucasian, African American, Hispanic, and Asian, ages ranging between 0 and 18, collected from Children's Hospital Los Angeles (CHLA). To increase the validity of the experimental results, a large set of radiographs from the mentioned database were used in experiments to dampen down the possibilities of low accuracy and skewing results due to extreme data. Besides, all experiments adopted hand bone radiographs which range from different ages, genders and races to prevent problems arising from incomplete data and bias data selection. Furthermore, the standard database is available online to enable other researchers repeat the conducted experiments using the same set of data to facilitate future comparisons. All the experiments in assessing computational algorithms were conducted using identical version of software and in identical personal computer to minimize random errors. Qualitative and Quantitative experiments were set up to justify the previously discussed arguments in this book:

(1) The first experiment was set up to justify the motivation of equalizing the image before the system and to justify the strengths of the proposed Multiple Beta Optimized Histogram Equalization (MBOHE) in comparison to other current histogram equalizations.

Y. C. Hum, *Segmentation of Hand Bone for Bone Age Assessment*,
SpringerBriefs in Applied Sciences and Technology, DOI: 10.1007/978-981-4451-66-6_4,
© The Author(s) 2013

(2) The second experiment was set up to justify the positive effects of adopting anisotropic diffusion in the segmentation framework compared to conventional filtering methods in tackling the problem of inherent high variations within the bone structures.
(3) The third experiments consisted of analytical evaluation and empirical evaluation was set up to justify the performance of the proposed segmentation framework.

4.2 Anisotropic Diffusion in the Proposed Segmentation Framework

In previous chapter, the details of anisotropic diffusion and its role in the segmentation framework have been explained. In this chapter, the effect of anisotropic diffusion on hand bone radiographs is justified. As shown in the Fig. 4.1, the bone area had been smoothed to become homogenous area; the black holes and dots in had been filled by similar pixel intensity with the surrounding bone. Despite this filtering process, the edges of hand structure are preserved and can be clearly seen.

Figure 4.2 illustrated the diffusion effects of various filtering methods: Gaussian filter, average filter, wiener filter [1] and Symmetric Nearest Neighbor (SNN) filter [2] and anisotropic diffusion. From this figure, it is observable that the effect of anisotropic diffusion was more favorable in comparison to other diffusion methods as the edges had been preserved and even enhanced while the degree of heterogeneity within the hand bones has been reduced by eliminating the random noise. Unlike anisotropic diffusion, the other filters were not adaptive to the information contained in the moving kernel. In other words, the filters imposed identical degree of diffusion and identical direction of diffusion to every single pixel without considering the suitability of information resided in the kernel. Therefore, as expected, the edges are blurred.

To quantify the changes in homogeneity, an experiment is conducted to evaluate the homogeneity of 12 radiographs from each age group after being diffused by anisotropic diffusion. Table 4.1 showed the experimental result of expected image homogeneity before and after the anisotropic diffusion processing. From the table, it is observable that the homogeneities after the diffusion are consistently larger the image homogeneities before anisotropic diffusion over all age groups. The increase of homogeneities is important to subsequent steps in the proposed framework's sub-routine. The homogeneities blend together the noises and smooth the uneven texture of bones so that the exhibited histograms of bone regions are more prone to having separated 'hills' that are inherently easier to be distinguished by relatively simple clustering or thresholding techniques.

As a whole, anisotropic diffusion has an edge over the traditional scale-space filtering methods in terms of the relatively low complexity in computation. This advantage capable to extend its applicability of anisotropic diffusion for general purposes; anisotropic diffusion takes structures outlines into consideration in filtering; therefore,

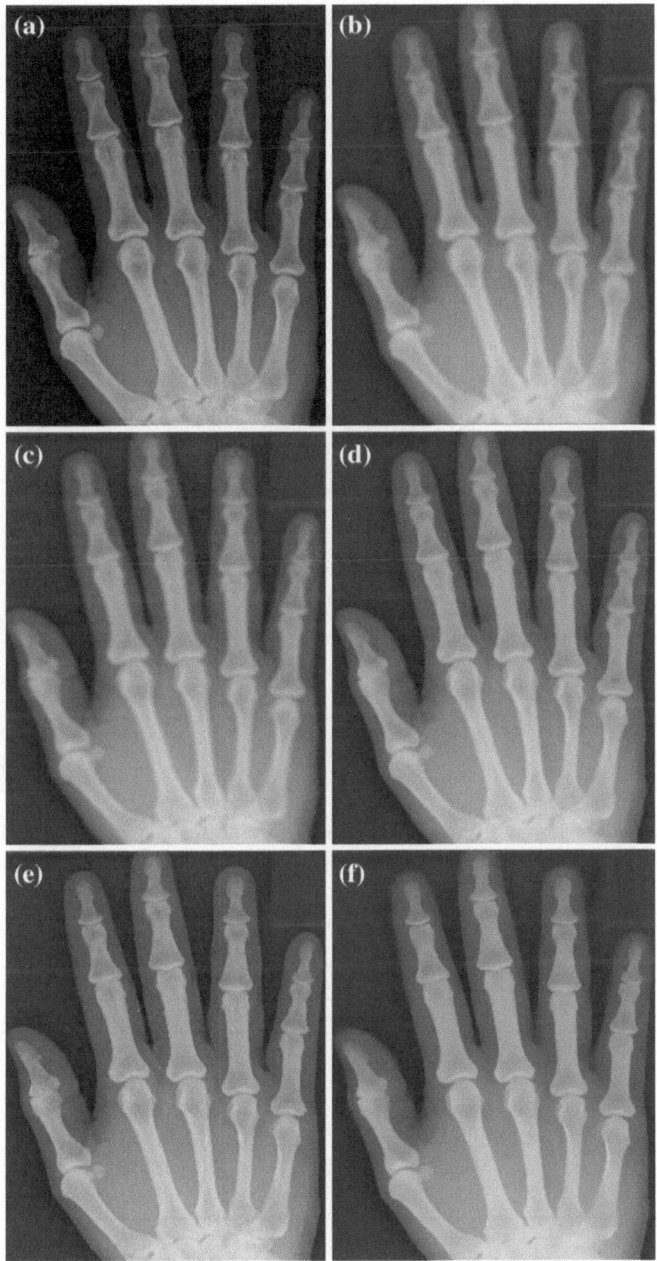

Fig. 4.1 An arbitrary *left* hand bone radiograph is used to demonstrate the anisotropic diffusion effect. **a** Hand radiograph Image *before* diffusion. **b** Hand radiograph Image *after* diffusion

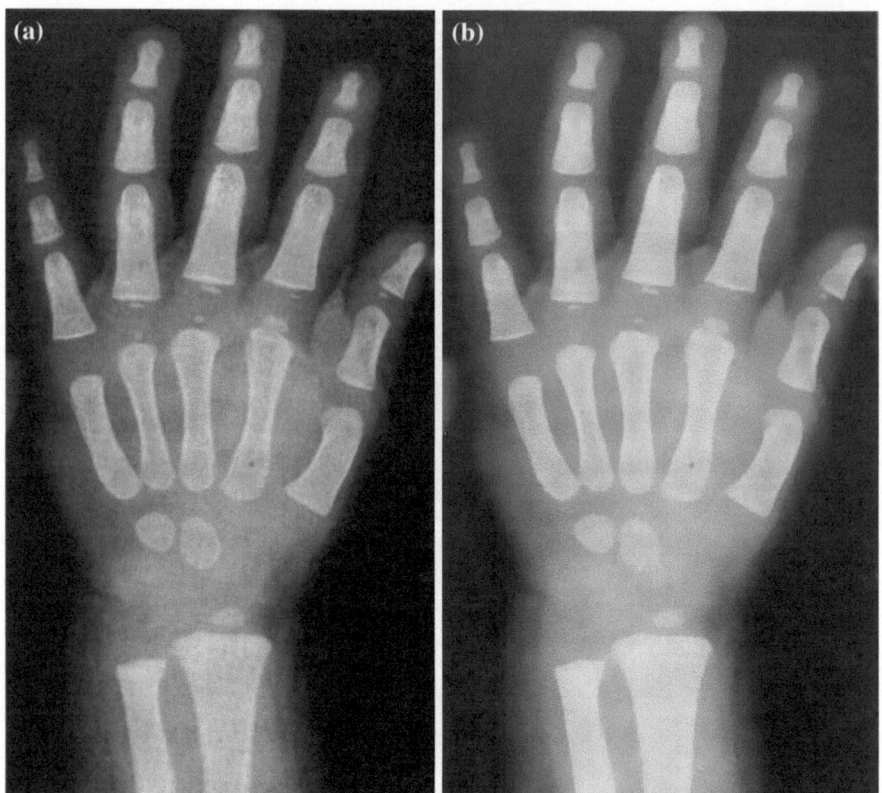

Fig. 4.2 The radiograph is diffused by different algorithm for comparison. **a** Original image. **b** Gaussian filter. **c** Average filter. **d** Wiener filter. **e** Symmetric Nearest Neighbor (*SNN*) filter. **f** Anisotropic diffusion

Table 4.1 Comparison of expected image homogeneity of different age group before and after the anisotropic diffusion processing

Age group	Image homogeneity	Image homogeneity
	Before anisotropic diffusion	After anisotropic diffusion
0–2	0.7056	0.8235
3–6	0.6984	0.8245
7–9	0.7132	0.8365
10–12	0.7189	0.8423
13–15	0.7028	0.8212
16–18	0.6927	0.8234

the edge has not been diffused and thus details are retained. Edges of structures, thus, are intensified to facilitate identification. Moreover, the anisotropic diffusion is able to manipulate the intensity and direction of diffusion to prevent diffusion across the edges; instead it enables diffusion in direction parallel along the edges; therefore, edges are preserved and are enhanced simultaneously. This is important in distinguishing the anatomical structures outlines in medical image processing.

4.3 Segmentation Evaluation

The segmentation evaluations methods can be categorized into analytical evaluation methods and empirical evaluation methods [3]. Which method is to be used in the segmentation evaluation depends on the segmentation algorithms that are to be evaluated. For example, if an enhanced version of a particular segmentation algorithm has been designed by changing one of its parameters, or the selection method for that particular parameter, then empirical evaluation method is more suitable because both segmentation methods contain the same amounts of parameters and thus it is possible to set all the other parameters of the particular segmentation as constant variables, the performance of segmentation as response variable and the modified parameter as manipulated variable. Thus, empirical segmentation evaluation can examine the effect of the particular modified parameter.

If segmentation algorithms are to be compared from different categories that are significantly different from each other in terms of number of parameters, type of parameters, parameters selection, number of inputs, types of inputs, and fundamental concept, then analytical methods that analyzes the, properties, applicability, limitations, utilities, computational complexity, and principles of those segmentation algorithms were more suitable. Besides being more suitable, the analytical evaluation methods require less concrete realization of algorithm implementations and this prevents the confrontation of experimental errors associate with setting up the experiments as in empirical evaluation methods. The analytical evaluation methods, nonetheless, has limitation because not all segmentation methods is analytically evaluable due to the scarcity of general segmentation theory.

In this chapter, comparison is made between the segmentation framework and the state-of-the-art AAM which is the well-known deformable model developed from ASM that have gained much attention in the image processing community. This technique has been specifically applied in hand bone segmentation and is still actively being applied at the moment. Both the details of ASM and AAM in the chapter of literature review have been explained. The second reason to compare with this method due to the fact that this is the latest segmentation method adopted by the only commercialized computer-aided skeletal age scoring system corporate known as 'BoneXpert' in the world thus far recently, to the best of knowledge. Besides, unlike other arbitrarily found techniques in literatures, the authors have included relatively substantial amounts of details of the techniques that would increase the reliability and the repeatability of the comparison and conducted experiments.

4.3.1 User-Specified Parameters

Segmentation algorithms are examined and compared using both analytical and empirical evaluation methods. In analytical evaluation, the listed properties of segmentation algorithms that are critical to practical usage of segmentation in computer-aided skeletal age scoring system are examined: by analysis of type, number and nature of user-specified parameters: The number of parameters that is required to be manually set by operators in order to execute the segmentation method. Via this analysis, the related segmentation properties are as follow:

a. Automaticity: The ability to complete the task of segmentation without or with least human intervention.
b. Adaptability: The ability to adapt to different types of content in hand bone radiographs with different number of ossification sites from categories of age 0 to 17.
c. Repeatability: The ability to produce similar performance outcomes (segmented image) in repeated executions, either by the same operators (intra-operator agreement) or by different operators (inter-operator agreement).

In empirical evaluation that involves implementation of algorithms on test images to assess the performance of algorithm, the accuracy is examined. The accuracy is defined as the ability to produce performance outcomes (segmented image) as close as possible to the ideal performance outcomes (segmented image) of segmentation algorithms. The accuracy is critical practical usage of segmentation in computer-aided skeletal age scoring system.

4.3.1.1 Active Appearance Model

The AAM hand bone segmentation framework by Thodberg et al. [4] used 1,559 training image, 3 epochs, 64 shape points, 12 principal shape models, 1,600 sampling points in intensity model, 30 principal intensity models, 4 pose parameters, 45 AAM models are trained despite the assumptions as follow:

(1) The number of epochs that represents each type of bone is fixed in each training sample.
(2) The number of divisions is fixed in each training sample in each training sample.
(3) The number of AAM is fixed in each training sample in each training sample.
(4) The number of principal shape model is fixed in each training sample.
(5) The intensity model is fixed in each training sample.
(6) The number of pose parameter is fixed in each training sample.
(7) The type of trained AAM to be fitted on hand radiograph during execution phase.
(8) The location of trained AAM on hand radiograph during execution phase.
(9) The number of model iterations is fixed in each hand radiograph during execution phase.

The total number of parameters that is still required to specifically be determined from training phase to execution phase for one hand radiograph segmentation amounted to 1,01,348. This is because each training image that contains different number of bones in different age group requires expert to determine the number of shape points, and each shape point, in AAM, is a parameter. Furthermore, each shape point required human operator to manually mark each of the location. Undoubtedly, each shape point location in this context is considered as a parameter as well. One should notice that the enormous number of decisions required to be made by human expert had not included the type of model, location of model and number of iteration to be determined during the execution phase because it has been assumed to be fixed which is not practical and not applicable. Therefore, generally, if it is assumed that only assumptions 1–6 above are acceptable, the total number of user-specified parameters N_{TNP} that is practical can be roughly be computed as follows:

$$N_{TNP} = \sum_{i=1}^{N_{TI}} (N_{SP}(i)) + [(N_{SP}(i)) + (N_{SPL}(i))] + 3(N_R) + 10 \quad (4.1)$$

where $N_{SP}(i)$ denotes the total number of shape points in ith training sample; $N_{SPL}(i)$ denotes the ith shape points location of $N_{SP}(i)$; N_{TI} denotes the total number of training samples, N_R denotes the total number of targeted hand radiographs to be segmented for computer-aided skeletal age scoring system. According to Eq. (1.1), if 100 hand bones from 100 radiographs are to be segmented, and the total shape points in all the training samples are identical, then the total user-specified parameters amounts to 3, 99,112. The type of parameters and number of parameters of each type is shown in Table 4.2.

Table 4.2 User-specified parameters used in shape-driven AAM

Type of parameters	Parameters
Training phase	
(1) Number of bones to model	1
(2) Number of epochs to represent the age range	1
(3) Division of epochs	1
(4) Number of AAM	1
(5) Number of training images	1
(6) Number of shape points, n	n for each training sample
(7) Locations of shape points	1 for each shape point for each training sample
(8) Number of principal shape models	1
(9) Number of sampling point in intensity model	1
(10) Number of principal intensity models	1
(11) Number of pose parameters	1
(12) The number of iterations before the trained AAM model fit to the hand bone	1
Execution phase	
(1) Choose which of the trained AAM models is to be fitted on the targeted hand bone	1 for each executed radiograph
(2) Place the location of the chosen trained AAM model	1 for each executed radiograph
(3) The number of iterations before the trained AAM model fit to the hand bone on each radiograph	1 for each executed radiograph
Total parameters (100 executed radiographs)	3,99,112

Similarly, the number of parameters ASM schemes used by various researchers [5, 6] can be computed using Eq. (4.1) with slight modification to exclude the two parameters involving intensity model in AAM as follows:

$$N_{TNP} = \sum_{i=1}^{N_{TI}} (N_{SP}(i)) + [(N_{SP}(i)) + (N_{SPL}(i))] + 3(N_R) + 8 \qquad (4.2)$$

4.3.1.2 The Proposed Framework

The type and number of user-specified parameters of the proposed segmentation framework are as shown in Table 4.3. Let N_{TNP} denotes the total user-specified parameters which can be analytically summarized as follows:

$$N_{TNP} = N_{iq} + 3 \sum_{i=1}^{N_{iq}} \left(N_{mf}(i)\right) + 8 \qquad (4.3)$$

Table 4.3 User-specified parameters used in proposed segmentation framework

Type of parameters	Parameters
MBOBHE	
a. Modeling phase	
(1) Relative weight of NBPS	1
(2) Relative weight of NOCS	1
(3) Relative weight of NDPS	1
b. Execution phase	
–	0
Modified anisotropic diffusion	
(4) The filter's kernel size	1
ACR segmentation	
(5) GLCM filter's kernel size	1
(6) Number of gray scale level	1
Fuzzy quadruple block division	
a. Modeling phase	
(7) Number of input variables, N_{iq}	1
(8) Number of membership functions within each input variables, $N_{mf}(i)$	1 for each input variable $= N_{iq}$
(9) Shape of each membership functions within each input variables, $N_{mf}(i,j)$	1 for each membership function $= \sum_i (N_{mf}(i))$
(10) Exact location of prototype of each membership function	1 for each membership function $= \sum_i (N_{mf}(i))$
(11) Defuzzification method	1
b. Execution phase	
–	0
BARNAE	
(12) Kernel size during entropy edge detection	1
Total parameter (100 executed radiographs)	42

where N_{iq} denotes total number of input variables; $N_{mf}(i)$ denotes the number of membership function in ith input variables. In the case of proposed segmentation framework, in accordance to Eq. (1.3), the total user-specified parameter amounts to 42.

Note that in Eq. (4.3), it does not contain the total number of training samples [the quantity denoted as N_{TI} in Eq. (4.1)] and the total number of targeted hand radiographs to be segmented for computer-aided skeletal age scoring system [the quantity denoted as N_R in Eq. (4.1)]. Another way of explaining is that the number of user-specified parameters will not increase with the number of radiographs that are to be segmented and it does not depends on any training samples because the modeling phase needs only to be accomplished once by the human expert to instill the intelligence and flexibility.

4.3.1.3 Interpretations

Figure 4.3 showed that the number parameters of the proposed segmentation framework are invariant to the increase of the number of targeted radiographs. As illustrated in the figure, the number of user-specified parameters increases linearly against the number of radiographs that are to be segmented (targeted radiographs) whereas the number of user-specified parameters of proposed segmentation framework remains unchanged against the number of targeted radiographs.

This criterion is of exceptionally importance when the objective is to segment a large amount of hand bones for computerized BAA. It means even if both segmentation frameworks are assumed to have similar number of parameters, the AAM segmentation framework is confronting with large amounts of parameters (linear increment) as the number of targeted radiographs increases. Besides, increment in user-specified parameters can be viewed as a sign of computational complexity increments. The reason is that if the intention is to replace all the user-specified parameters decision using automated methods (if exist), higher level of image

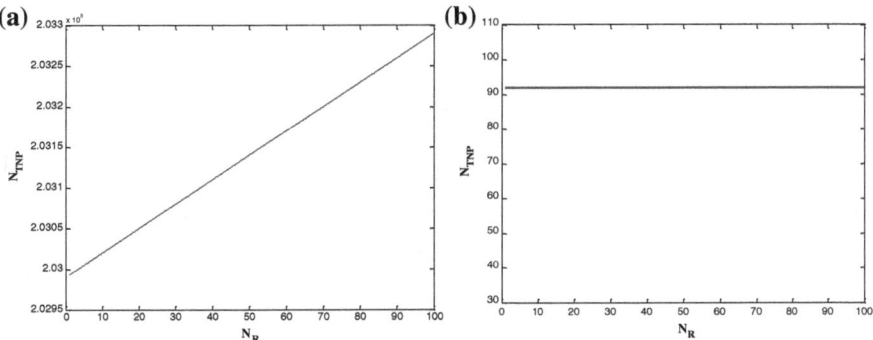

Fig. 4.3 The change of total number of parameters against the increase of targeted radiographs for **a** AAM. **b** Proposed segmentation framework

analysis procedures are required, and this definitely increases the computational efficiency if the number of user-specified parameters increases. Similarly, the number of training samples, N_{TI} exhibits linear incremental growth in total number of user-specified parameters while the proposed method are not affected by training samples.

The influence of the number of targeted radiographs and training samples to the total number of user-specified parameters has been analyzed. Now the influence of number of shape points and the number of shape point locations to the total number of user-specified parameters are analyzed. Obviously, both number of shape points and number of their locations are of the same amounts, therefore, $(N_{SP}) = (N_{SPL})$, (N_{SPL}) contains 2 parameters, which are the 2D coordinations of the shape point location, and thus Eq. (1.2) is redefined to Eq. (1.4). To examine the influence, the other factors such as N_R and N_{TI} are assumed to be the constants, C, as shown in Eq. (1.5). Figure 4.4 illustrates that the N_{TNP} exhibits linear-like growth as N_{SP} increases on 100 targeted radiograph using 1,559 training samples. This indicates that if the number of shape points is increased in order to capture the complicated shape of hand bones when the patients ages increases, the number of parameters that requires expert to insert increases linearly.

$$N_{TNP} = \sum_{i=1}^{N_{TI}} 4\,(N_{SP}\,(i)) + 3(N_R) + 10 \qquad (4.4)$$

$$N_{TNP} = C\,(N_{SP}\,(i)) + C \qquad (4.5)$$

The increment of the parameters such as the number of membership functions, the shape of the membership functions, and the location of the membership functions also exhibit the similar linear growth effect to the N_{TNP} similar to Fig. 4.4. This indicates only that, for example, if the number of membership function increases, the N_{TNP} grows with same pace as in N_{SP}. However, the gist of the

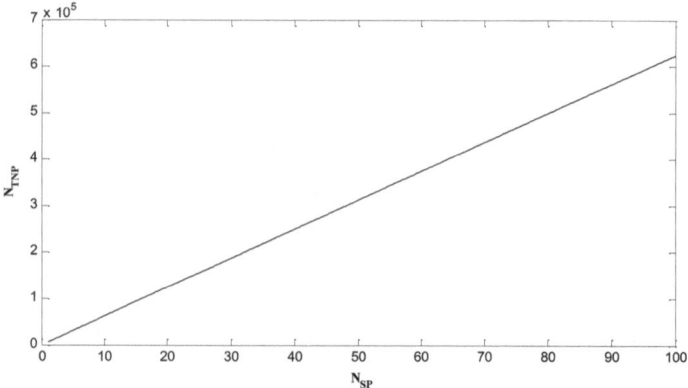

Fig. 4.4 The change in N_{TNP} as N_{SP} increases

problem is that N_{SP} always requires a substantial numbers so that to sufficiently describe the shape of the hand bone, for example, in [4], the author utilized 64 shape points to delineate the hand bone shape. On the contrary, in the fuzzy quadruple division scheme, or any fuzzy inference system, the number of membership functions is averagely around 3–6 in order to sufficiently describe the expert vague knowledge. This explanation goes similarly to the training examples; in [4], the author utilized 1,559 samples to train the model so that the model could capture the changes of shapes existed in the samples. However, in the proposed segmentation framework, no training sample is required. The large numbers of shape points and training samples magnify the constants in the Eq. (4.5) as shown in Eq. (4.6). In contrast, it is compared to Eq. (4.7), which is the numerical representation of Eq. (4.3):

$$N_{TNP} = (1559 * 4)\,(64) + 10 = 3,99,112 \tag{4.6}$$

$$N_{TNP} = 4 + 3\,[(2) + (2) + (3) + (3)] + 8 = 42 \tag{4.7}$$

The above illustration explains the reason for the large number of difference between the N_{TNP} of AAM and the proposed segmentation framework. From the comparisons and arguments above, the impact of the user-specified parameters for both AAM and the proposed segmentation framework in hand bone segmentation in following paragraphs is interpreted.

The proposed segmentation framework requires relatively much less numbers of user-specified parameters if it is compared to AAM framework. The above fact can be viewed from a few perspectives. First of all, from the perspective of automaticity, large number of user-specified parameters simply means that it needs human decisions and labors to perform the task; the higher the number of user-specified parameters indicates the higher the difficulty in transforming those parameters becoming automated. From the perspective of the time taken in segmentation task, the high number of user-specified parameters indicates high time consumption as each parameters needs to be adjusted. From the perspective of adaptability, higher number of user-specified parameters can indicate higher practical flexibility and adjustment because the operator possesses powerful cognitive ability relative to computational algorithm. From the perspective of noise resistance, higher number of user-specified parameters in the context of AAM in hand bone segmentation can be considered as being inherently more robust because the trained model has been placed very close to the expected location of each structure in hand bone and thus the trained model is less susceptible to the noises in soft-tissue region and background (the proposed segmentation framework has incorporated the anisotropic diffusion to reduce the influent of noises found bone structure, soft-tissue region and background).

The types of the parameters of both frameworks are different. Thus, the natures of those parameters are different as well. The user-specified parameters in proposed segmentation framework can be easily replaced by simple analysis or assumptions such as in the relative importance of the MBOBHE, the relative importance of the three properties of histogram equalization can be assumed as being equal. Even if the result is not optimized, the difference is not obvious. Besides parameters such as the shape of the prototype in membership functions have been proven that minor deviations from the ideal setting are not influential enough to deteriorate the final result. Nonetheless, the user-specified parameters in AAM such as the shape points placements and shape points numbers are very specific. A wrong placement of shape point, insufficient number of shape points and underestimated iterations can be disastrous to the final result; therefore, the potential of automating these parameters is reduced by the nature of the parameter. From the perspective of noise resistance, having a parameter that requires specific and accurate tuning is not favorable because it overly depends on certain feature that could possibly deteriorate the final result if the feature is not as ideal as the algorithm's expectation (relates to the P9 segmentation criterion in previous chapter).

The proposed segmentation framework does not require training radiographs whereas the AAM framework requires a large amounts training radiographs and shapes points. This criteria of require no training image is one of the most important feature of the proposed segmentation framework. Even if the set of hand bone radiographs sample ranging from age 0 to 17 is incomplete, the segmentation framework can be performed and the result has no different whether it has more or less samples of radiographs. This contributes to the applicability of the segmentation framework that it can be applied directly in any hospital without demanding a large set of radiographs samples. In other words, lacking of training samples will not influent the efficacy of the proposed segmentation framework, but contrarily will impose profound effect on AAM segmentation framework because the model simply could not be trained or the shape of the desired anatomical structures could not be captured.

The number of user-specified parameters in proposed segmentation framework is invariant to the number of targeted radiographs whereas the number of user-specified parameters in AAM framework will grow linearly with the increment of the number of targeted radiographs. From the perspective of applicability, the efficiency of AAM deteriorates as the number of targeted radiographs increases because the number of user-specified parameters increases and it means that the operators have to make decisions on a large number of parameters which could result in high error rate. The proposed segmentation framework requires no user-specified parameters during segmentation execution phase, compared to AAM framework that requires 3 user-specified parameters during execution phase. Such 3 user-specified parameters are, technically, difficult to be made automated for the users. Thus, those decisions such as the choice of trained model, the location to place the model, and the iterations before the model fits to the hand bone, require expert user equipped with knowledge in both bone age assessment and image processing. In other words, the applicability of the computer-aided skeletal age scoring system that adopts such a segmentation framework is limited to the presence of expert users. On the contrary, the proposed segmentation framework can be operated by any layman user because the framework has been made autonomous with expert knowledge. This property of automaticity is so critical that it enables the realization of perfect repeatability without intra-operator deviation and without inter-operator deviation.

4.3.2 Segmentation Accuracy

It is useful to evaluate the accuracy of the proposed segmentation framework on hand bone segmentation. The purpose is to examine, despite having less user-specified parameters, in a newly designed framework, can the accuracy of segmentation achieve the similar level to AAM. The accuracy of AAM though is difficult to be measured as there are many undefined parameters in literatures of hand bone segmentation using this method. Secondly, it is obvious from the theoretical aspects of AAM that only if it is given sufficient number of training images that encompasses sufficient expected changes of hand bone number, locations and sizes, all parameters are correctly tuned, hand bone anatomical structures of all training radiographs are accurately delineated, only then the segmentation can be almost perfect. Note that not all segmentation methods can hit the accuracy of being 'almost perfect' even if the parameters are given the best tunings; for example, the global thresholding method discussed in Chap. 2; even the 'best' threshold is chosen, the threshold is unable to overcome the uneven luminance and the inherent intensity distribution of hand bone radiographs. Therefore, it was the objective of this sub-section to justify the accuracy of the segmentation framework, to examine its 'best' accuracy, despite having less user-specified parameters.

There are multitude segmentation techniques and frameworks exist in literature [3, 7, 8]. However, difficulties arose if the performance of those developed segmentation methods in literature are to be measured. It is claimed that the unsupervised segmentation evaluation has an edge over other type of segmentation evaluation such as unsupervised segmentation evaluation and human visual inspection segmentation evaluation. The key advantage is its self-tuning potential since no reference image is needed as in supervised evaluation. Besides unsupervised evaluation, there is another common group of methods in segmentation evaluation which can be categorized as supervised evaluation. Supervised evaluation is a comparison made against a predetermined manually segmented reference image. Similar to human inspection method, supervised evaluation involves human assistance in the early stage such as segmenting a reference image, which could be even more tedious than human visual inspection alone, especially in the context of evaluating hand bone segmentation; this is due to the inherent countless variations in number and size of ossification sites along the hand bone development.

Attributable to the abovementioned reasoning, unsupervised methods have been adopted in this book as performance evaluation to assess some of the result such as in evaluating the abilities of histogram equalization, firstly, in solving the problems of over-enhancement, detail and brightness preservation as well as the contrast enhancement; secondly, in evaluating the ability of anisotropic diffusion in solving the problem of high diversity in pixel intensity in hand bone radiograph. However, unsupervised evaluation is not adopted in evaluating the segmentation result in hand bone segmentation. Unsupervised evaluation of segmentation in literatures, though have an edge in term of time taken and objectivity over human visual system, they lack of background knowledge as in human and they are too general to be applied in such a specific system such as computer-aided skeletal age scoring system which requires evaluator to distinguish which features are pertinent in segmented image that determines an accurate result of bone age assessment.

A so-called 'good' segmentation evaluation is a relative quality. Whether it is favorable should depend on the application. As each application has different focus and hence different criteria should be categorized as a favorable resultant segmentation. In the context of bone age assessment, the focus is on the features which will affect the score in bone maturity such as the thickened white line along the border, continuous border of epiphyseal center, the diameter ratio of the epiphyseal centre to the metaphysis, the dark line of cartilage, distinct thickened proximal border and etc. However, to the best of our knowledge, the existing unsupervised segmentation evaluation methods are not tailored for hand bone segmentation in the application of computerized TW3 bone age assessment. Using a general segmentation evaluation is inappropriate and inaccurate to reveal the real performance of each segmentation method.

Due to the above reasons, a specially tailored human visual based evaluation method for hand bone segmentation in the context of bone age assessment is designed and used to evaluate the segmentation result. This type of supervised evaluation can be categorized as subjective evaluation as well, which is claimed to be the most common evaluation methods. The major disadvantage of this subjective evaluation is that it is very subjective, as its name suggests. However, if the reference object such as hand bone is manually segmented with high accuracy and the relative contribution of each feature are set correctly, and then the evaluation can be regarded as considerably objective. Besides, it is very time-consuming if the experiments involve a large number of data. However, in the case of this book, such evaluation is the best evaluation compared to other existing supervised and unsupervised evaluation method. Therefore, to maximize its accuracy of evaluation and minimize the bias caused by subjective evaluation, a large set of test images from the database are adopted, and four evaluators who have been inspecting more than 1,500 hand bone radiographs for manual TW3 bone age assessment are given task to evaluate the result. Large set of test images and multiple evaluators are effective in reducing the undesired bias such as subjectivity. Another important factor is a set of well-designed, clear and application-related guidelines in the evaluation given to the evaluators.

A set of good guidelines should be related to the application or the purpose of the segmentation. In the context of bone age assessment, the key point is that if less concern is on evaluating how accurate the segmented image resemblance the desired segmented image for the bone areas, but how much pertinent information has been retained in the segmented image. An example is shown in Fig. 4.5 to illustrate this concept. Both Fig. 4.5a, b have same amount of pertinent information as the regions-of-interest (regions within red rectangular) are not affected by the artifacts, or in other words, the artifacts (region within blue rectangular) of Fig. 4.5b occur in non-region-of-interest. In terms of information preservation, both images are identical, in the context of TW3 bone age assessment.

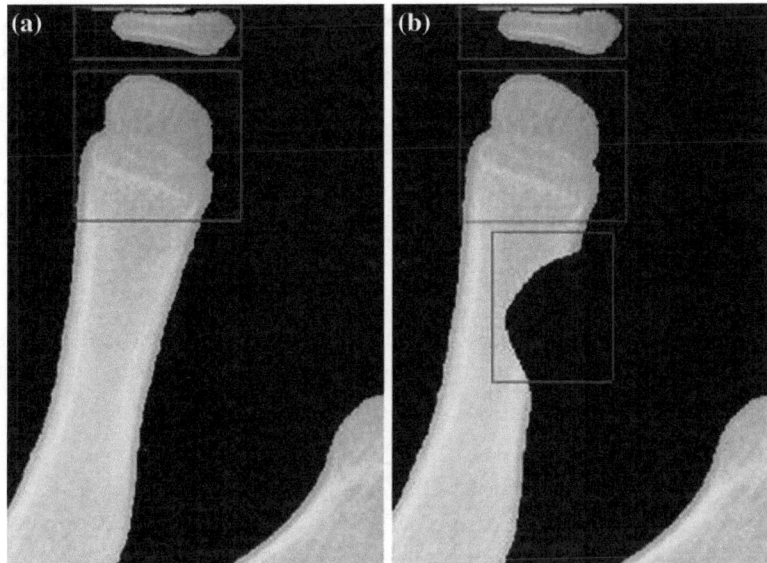

Fig. 4.5 **a** Block of segmented hand bone without any artifact. **b** Block of segmented hand bone with artifact in non-region-of-interest

However, in terms of potential value of usefulness for subsequent procedures in computerized bone age assessment, both images have distinct differences. Figure 4.5a has much more potential value of usefulness than Fig. 4.5b, as the artifacts in Fig. 4.5b will pose inferior effects on subsequent pattern recognition procedures especially in automated localization of regions-of-interest. In short, both pertinent information preservation and potential value of usefulness have to be taken in consideration when assessing the quality of hand bone segmentation; however, more concern should be given to pertinent information preservation as they are the fundamental bone age determination material.

After comparing various supervised evaluations [9, 10], three out of them that seem to be particularly appropriate to express the importance in the evaluation of the segmented hand bone in the context of computer-aided skeletal age scoring system application have been chosen: Figure Of Merits (FOM) proposed by [11], Figure Of Certainty (FOC) and Fragmentation (FRAG) proposed by [12]. One of the main reasons of choosing them is because of their generality; these metrics can be easily customized into specific evaluation context due to the scaling parameters in the metrics. Secondly, each of them is capable of measuring one of the important features in segmented hand bone in the context of computerized BAA using different weight to express the different relative importance.

The explanation begins with the segmentation quality metrics which are the functions that measure the degree of similarity between the segmented image by segmentation algorithms and the ideally segmented image (ground truth). The metric returns a numerical value after comparing the segmented image with the reference image that is manually segmented indicating the similarity between them to rank the segmentation algorithms performance. The ground truths are defined differently in different segmentation quality metrics; for FOM, the ground truth refers to the edges of hand bones; for FOC, the ground truth refer to the pixel intensity of hand bones; for FRAG, the ground truth refers to the number of isolated fragmented bones resides in hand bone radiographs. The details of each metric are explained below:

(1) Figure of Merits (FOM)

It is the commonly adopted edge detection evaluation that measures the discrepancy between the detected edge pixels and the actual edge pixels. The metric is defined as follows:

$$FOM = \frac{1}{N} \sum_{i=1}^{N} \frac{1}{1 + \omega[E(i)]} \tag{4.8}$$

where N denotes the maximum number between the detected edge pixels and actual edge pixels. $E(i)$ denotes the distance between the ith detected edge pixel with its according actual spatial location. The parameter ω denotes the scaling parameter that manipulates the influence of errors to the metric; this scaling enables us to model the relative importance of edges in ossification sites and edges in non-ossification sites.

(2) Figure of Certainty (FOC)

It is a metric that measures the discrepancy of the gray level intensity of specific spatial location between the segmented hand bone and the gray level intensity in reference hand bone. The value reflects the degree of false labeling. Specifically, false labeling occurs if a pixel that constitutes the bone region is labeled as background pixel or if a pixel that constitutes soft-tissue region or background pixel is labeled as bone region. The FOC is defined as follows:

$$FOC = \frac{1}{N} \sum_{i=1}^{N} \frac{1}{1 + \alpha |R(i) - S(i)|^{\beta}} \tag{4.9}$$

where N denotes the total number of hand bone pixels. $R(i)$ refers to the ith pixel of hand bone in reference hand bone. $S(i)$ refers to the ith pixel of hand bone in segmented hand bone. The scaling parameter α determines the impact of small deviation between the reference pixels and segmented pixels whereas the scaling parameter β determines the impact of large deviations between the reference pixels and segmented pixels.

(3) Fragmentation (FRAG)

For accurate bone age assessment, the condition in which the total number of bones in actual hand bone radiograph amounts to total number of bones in segmented hand bone radiograph is crucial. A metric that is used to measure the extent of this agreement expresses the discrepancy level of bones number.

$$FRAG = \frac{1}{1 + \gamma \, |n_S - n_A|^\delta} \tag{4.10}$$

where n_S denotes the number of bones in segmented hand bone radiograph, n_A denotes the number of bones in actual hand bone radiograph, α denotes the parameter γ that determines the impact of small discrepancy between n_S and n_A whereas the parameter δ determines the impact of large discrepancy between n_S and n_A. If a portion of bone is missing, the observer has to decide visually whether the bone has more than 70 % of the desired region of actual bone to consider that it has been segmented.

Eight hand bone radiographs from each age group ranging from 0 to 18 in dataset have been tested in which two radiographs are chosen from each population of four populations in each age group with different gender. Therefore, the experiment comprised of 144 radiographs in dataset. This experiment adopted the mean value of 3 metrics above to quantitatively approximate the evaluation of the segmentation accuracy with appropriate fixed parameters values: $\omega = 0.8$ if the edge pixels in actual image are belong to ossification sites, else $\omega = 1$; $\alpha = 0.8$; $\beta = 0.9$ if the pixels are belong to ossification sites, else $\alpha = 0.6$, $\beta = 0.8$; $\gamma = 0.9$; $\delta = 1$ if the regions of bones are belong to ossification sites, else $\gamma = 0.3$; $\delta = 0.5$. The values of 3 metrics, $\overline{FOM}$, $\overline{FOC}$ and $\overline{FRAG}$ are computed as follows:

$$\overline{FOM} = \frac{1}{NM_j} \sum_{j=1}^{M_j} \sum_{i=1}^{N} \frac{1}{1 + \omega[E(i, j)]}, \quad j = 1,2,3 \ldots 18 \tag{4.11}$$

where M_j denotes the total number of radiographs in jth age group; $E(i, m_j)$ denotes distance between the ith detected edge pixel with its according actual spatial location of radiograph from jth age group.

$$\overline{FOC} = \frac{1}{NM_j} \sum_{j=0}^{M_j} \sum_{i=1}^{N} \frac{1}{1 + \alpha \, |R(i, j) - S(i, j)|^\beta} \tag{4.12}$$

where M_j denotes the total number of radiographs in jth age group; $R(i, j)$ denotes the ith actual pixel grey level intensity of radiograph from jth age group; $S(i, j)$ denotes the ith segmented pixel grey level intensity of radiograph from jth age group.

$$\overline{FRAG} = \sum_{j=1}^{M_j} \frac{1}{1 + \gamma \, |n_S(j) - n_A(j)|^\delta} \tag{4.13}$$

where M_j denotes the total number of radiographs in jth age group; where $n_S(j)$ denotes the number of bones in segmented hand bone radiograph of jth age group, $n_A(j)$ denotes the number of bones in actual hand bone radiograph of jth age group.

4.3.2.1 Evaluation on Automated Fuzzy Quadruple Division Scheme

The defined metrics have been implemented to evaluate the outcome of quadruple division scheme to justify its utility in the segmentation framework. Firstly, the qualitative assessment is conducted to visually observe the effect of the scheme by inspecting the segmented bone radiographs. Comparisons are made between the rigid adaptive division schemes with the proposed fuzzy quadruple division scheme on a quadrisected radiographs to inspect the differences. The input images are processed by the similar pre-processing in the proposed segmentation framework (constant variables), then they were implemented separately manually set rigid adaptive division scheme with different division layers and fuzzy quadruple division scheme (manipulated variables), the visual effects of those images are inspected and analyzed on qualitative analysis using human visual system and metrics are implemented to conduct the quantitative analysis on a large image sets to further justify the qualitative analysis (response variables). The visual segmentation result of TR-2.1 was shown in Fig. 4.6 to illustrate the differences.

Figure 4.6a denotes the original quadrisected radiograph; Fig. 4.6b denotes the TR-2.1 that was implemented with no division at all; Fig. 4.6c denotes the TR-2.1 that was implemented with division of 2 rows and 2 columns; Fig. 4.6d denotes the TR-2.1 that was implemented with division of 3 rows and 3 columns; Fig. 4.6e denotes the TR-2.1 that was implemented with division of 4 rows and 4 columns; Fig. 4.6f denotes the TR-2.1 that was implemented with the proposed automated fuzzy quadruple division scheme.

The effects illustrated in Fig. 4.6 were interpreted in a few perspectives. From the perspective of delineated bone structures, Fig. 4.6b–e failed to delineate the bone structures, meaningful outlines of hand bones could hardly been observed, however, the Fig. 4.6f that had undergone quadruple division scheme produced segmented hand bone that could at least provide noticeable hand bone shape. From the perspectives of segmented bone structures region, Fig. 4.6b–e showed that rigid division scheme produced excessive noises and failed to label correctly for most of the pertinent regions of ossification sites, whereas, Fig. 4.6f showed that via the proposed division scheme, most of the regions of desired bones structures were obtained. The reason lies in the adaptability of both scheme: the rigid scheme divides the radiograph regardless of the information that was contained inside the input and also regardless of the previous ACR algorithm; on the contrary, the superior performance of fuzzy quadruple scheme is due to the consideration on the information of the inputs and also the information of the output of previously implemented ACR algorithm to dictate the number and size of divisions. Note that all the images had not been processed by the quality assurance step; in next sub-section, the refinement of segmented hand bone shown in Fig. 4.6f is discussed using proposed quality assurance step.

Figure 4.6 showed only an arbitrary example of the processed radiograph to convey the qualitative visual effect of the proposed scheme compared to rigid

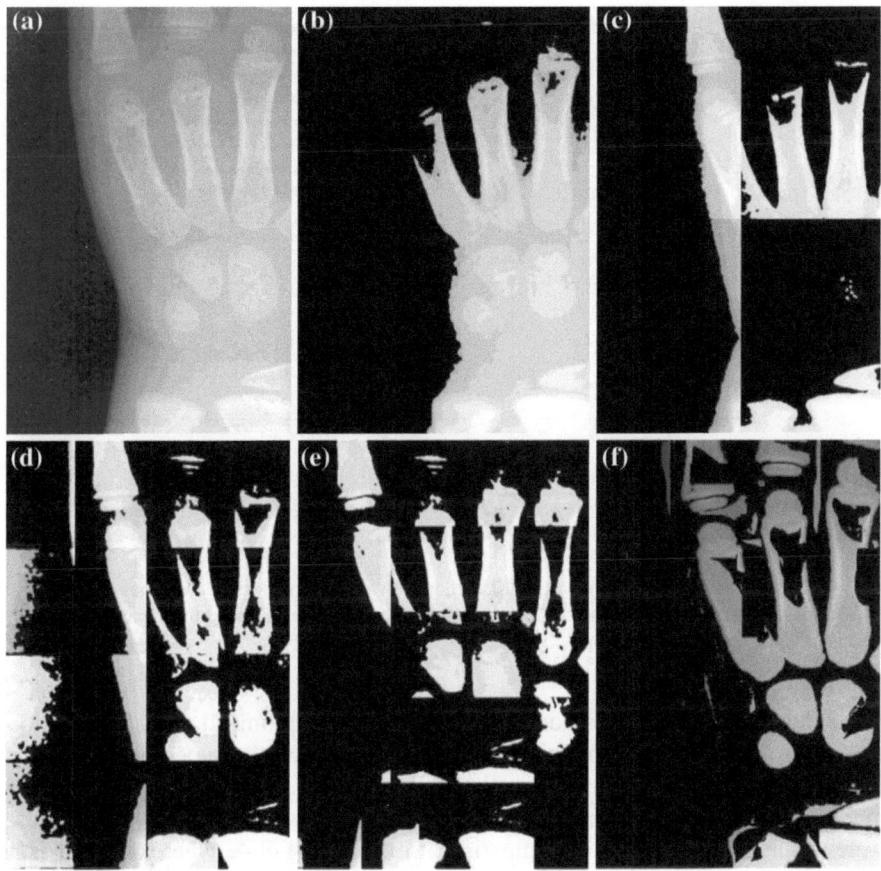

Fig. 4.6 Qualitative visual inspection on the differences between manual rigid adaptive schemes on an arbitrary TR-2.1

adaptive scheme. To further justify the effect statistically encompassing all the dataset, quantitative analysis is needed. As mentioned, we adopted the defined metrics of with specified parameters as described in Sect. 4.3.2 to quantify the predicted improvement. The results are plotted as shown in Fig. 4.7.

Figure 4.7 have shown obviously that the resultant segmentation accuracy of the proposed automated fuzzy quadruple scheme outperforms the other rigid division schemes in terms of correct labeling of edge pixels which are measured quantitatively by $\overline{FOM}$. The expected value across all age groups of the proposed scheme was 0.5721, followed by 0.3271 from rigid scheme of row = 4, column = 4, 0.2129 from rigid scheme of row = 2, column = 2, 0.2614 from rigid scheme of row = 3, column = 3, and 0.1614 from rigid scheme of row = 1, column = 1. The overall average of these rigid schemes in terms of $\overline{FOM}$ was 0.2407. Therefore, in terms of percentage, if compared to conventional rigid schemes, the proposed scheme improved by 118.85 %. This tremendous improvement over rigid scheme

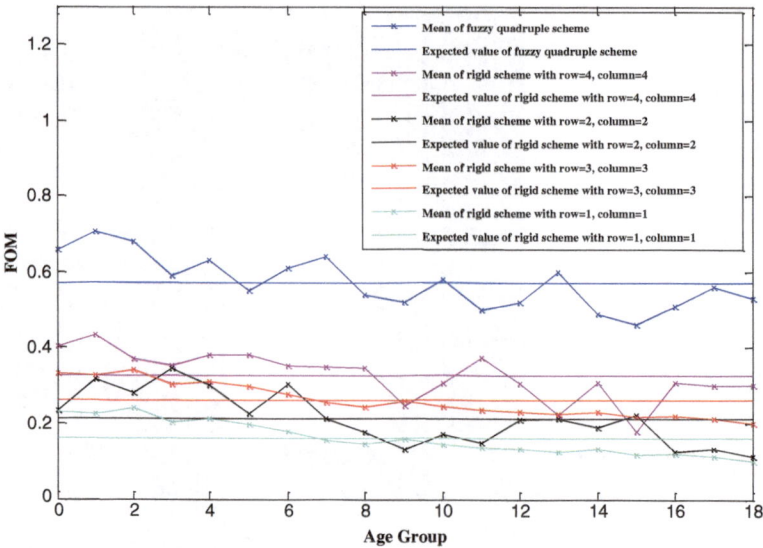

Fig. 4.7 Comparisons of segmentation accuracy between manual rigid schemes with various divisions and proposed automated fuzzy quadruple division scheme using the mean of FOM and expected value over all age group

was as expected. The proposed division scheme had optimized the adaptive block image size for ACR segmentation algorithm and hence the superior results were obtained. This result was intuitive according to the qualitative observation of segmented bone shown in Fig. 4.6 that most of the critical edges of bones structures were retained in Fig. 4.6f in comparison to the Fig. 4.6b–e where most of hand bone outlines were inferiorly segmented, most importantly, the edges of ossification sites that carried higher weights were missing.

Figure 4.8 illustrated the segmentation accuracy comparison between rigid division schemes and the proposed quadruple division scheme. The metric $\overline{FOC}$ quantified the degree of discrepancies in labeling of gray level between reference radiograph's pixels and the segmented radiographs over all age group among rigid division schemes and the proposed quadruple division scheme. The above experimental results have showed that the proposed division scheme possessed apparent superior performance of pixel labeling in comparison to other rigid division scheme. Specifically, the expected value of proposed division scheme over all age groups was 0.5179, followed by 0.2755 from rigid scheme of row = 4, column = 4, 0.2156 from rigid scheme of row = 2, column = 2, 0.1454 from rigid scheme of row = 3, column = 3, and 0.1906 from rigid scheme of row = 1, column = 1. The overall average of these rigid schemes in terms of $\overline{FOC}$ was 0.2068. In terms of percentage, averagely, the proposed scheme performed 150.4 % better than rigid schemes. Such an improvement was expected from the result of Fig. 4.7. This improvement indicated that the false labeling of proposed scheme in performing the ACR algorithm in the problem of hand bone

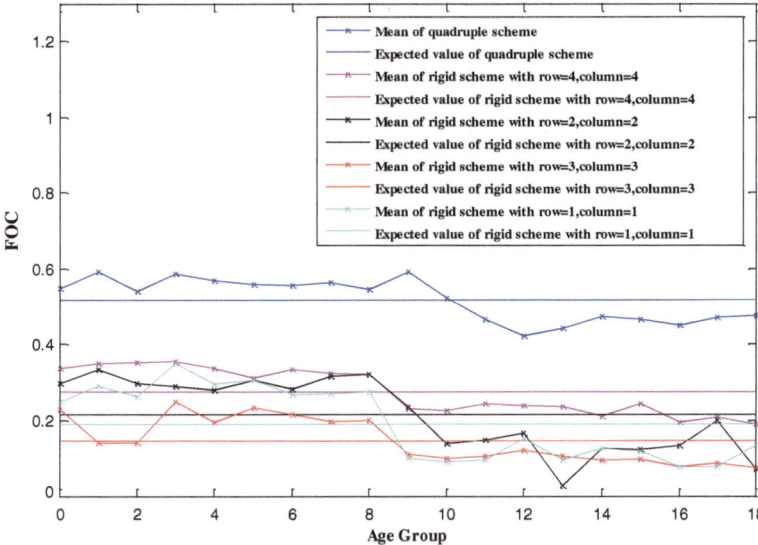

Fig. 4.8 Comparisons of segmentation accuracy between manual rigid schemes with various divisions and proposed automated fuzzy quadruple division scheme using the mean of FOC and expected value over all age group

segmentation was not trivial but noticeable enough to justify the claimed utility of the proposed scheme in retaining important regions of hand bones.

Figure 4.9 illustrated the comparison in terms of number of related bones that have been segmented. The improvement shown in result of $\overline{FOC}$ in above experiments compared to other rigid schemes was even more tremendous than the previous $\overline{FOM}$ and $\overline{FOC}$. This result revealed that even without the quality assurance process, the proposed division scheme is capable of retaining most of the critical type of bones that would affect the eventual bone age assessment result. Quantitatively, the average of $\overline{FRAG}$ over all age groups is 0.6997, followed by 0.1786 from rigid scheme of row = 4, column = 4, 0.1471 from rigid scheme of row = 2, column = 2, 0.1198 from rigid scheme of row = 3, column = 3, and 0.0844 from rigid scheme of row = 1, column = 1. The overall average of these rigid schemes in terms of $\overline{FRAG}$ was 0.1324. Note that the overall average of rigid scheme in this metrics are the lowest compared to the previous $\overline{FOM}$ and $\overline{FOC}$. This indicated that the rigid division scheme performed inferiorly especially in retaining the critical number of ossification sites. In terms of BAA, $\overline{FRAG}$ is considered to be more important than $\overline{FOM}$ and $\overline{FOC}$ because a single missing of ossification sites would totally distort the final result of BAA.

As a summary to this sub-section, experimental result on proposed division scheme with conventional rigid scheme, in a narrow sense, revealed that the proposed scheme did play a pivotal role in the segmentation framework and thus this justified the initial motivation of this scheme. In a broader sense, the result suggested the general insight that merely an explicit alternation of the properties for

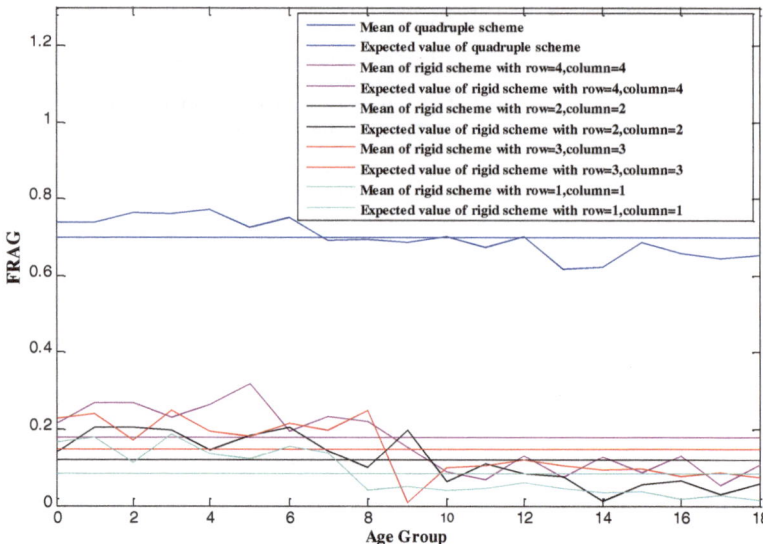

Fig. 4.9 Comparisons of segmentation accuracy between manual rigid schemes with various divisions and proposed automated fuzzy quadruple division scheme using the mean of FRAG and expected value over all age group

input (in specific context of this book, the property referred to the size of block radiograph) supplied to segmentation algorithm can be an effective alternative technique to improve the performance of segmentation algorithm instead of implicitly improving the segmentation algorithm itself.

4.3.2.2   Evaluation on Quality Assurance Process

As shown in Fig. 4.6f that even the diffused and equalized input block of hand bone radiograph that have undergone ACR algorithm in adaptive quadruple scheme failed to segment perfectly the desired hand bones; this is expected in light of the mentioned challenges and difficulties that reside in the segmentation techniques and the inherent properties of the radiographs itself and in fact, this is the motivation of the proposed segmentation framework. However, thus far, it is known that the most critical factor that influent the BAA, i.e., the number and type of bones that determine the eventual bone age according to TW3 using RUS, can be obtained via the previous processes from MBOBHE to the ACR algorithm in quadruple division scheme. Therefore, the quality assurance process was proposed and explained in last chapter that it functions as a filter that eliminates the undesired areas and fills in the desired regions. Generally, it acts as artifacts removers. This process is to enhance the segmentation accuracy automatically with a few steps including entropy—based edge detection together with the proposed BRANEA algorithm mentioned in last chapter. Therefore, in this chapter, the justification of the positive effect of this step to assess its utility in the segmentation framework is discussed. Firstly, qualitative evaluation is conducted by inspecting the visual effect of segmented radiograph under BRANEA algorithm. The comparisons were illustrated in Fig. 4.10.

This figure illustrated the effect BRANEA on segmented radiographs. Figure 4.10c or Fig. 4.6f represented the segmented bone regions without being post-processing by BRANEA, whereas Fig. 4.10e showed the final segmented bone region. It was noticeable that the unwanted of bone regions had been removed and the false labeling of bone region as background pixels have been recovered. Figure 4.10b showed the detected edge using entropy method described in last chapter. Note that the detected edge was not perfect and in fact there are a lot of bone outline have been missed out. This was expected since edge detection as described in Chap. 2 that it has a lot of drawbacks and inability in completely

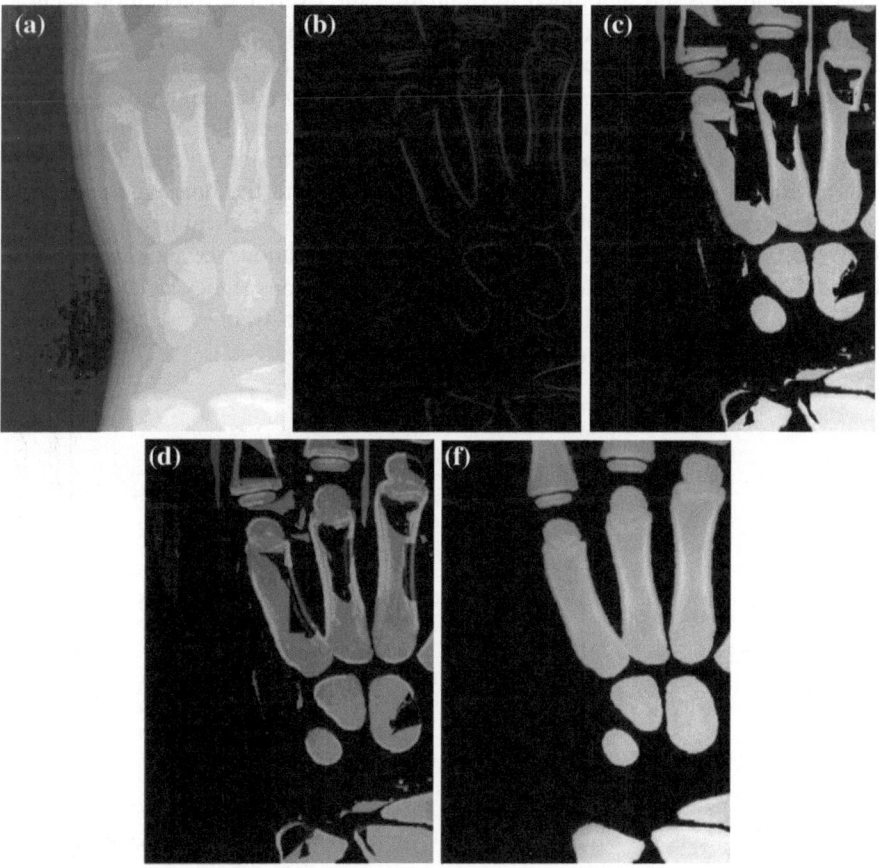

Fig. 4.10 Qualitative evaluation of BRANEA algorithm in recovering and eliminating artifacts. **a** Original. **b** Detected edge. **c** Segmented regions without BRANEA. **d** Combination of edge and segmented regions. **e** Final segmented region after BRANEA

delineating the anatomical structure; this was logical because if the edges of the
hand bone could have been completely detected, then it would not have needed
any segmentation techniques to partition the hand bone. Therefore, the Fig. 4.10c
provided the main regions and the detected edges played the role as supplemen-
tary material that complete the hand bone outline as shown in Fig. 4.10d. The
final segmented hand bone of Fig. 4.10e was obtained after applying BRANEA in
Fig. 4.10d.

4.3.2.3 Accuracy Evaluation of the Proposed Segmentation Framework

After inspecting the segmentation effect of BRANEA, quantitative evaluation
is conducted to further justify its consistency over all age groups of radiographs
by using $\overline{FOM}, \overline{FOC}$ and $\overline{FRAG}$ of metrics described in Eqs. (4.11–4.13)
This evaluation result amounted to evaluating the overall segmentation accuracy
of the entire proposed segmentation framework. The results were illustrated in
Figs. 4.11, 4.12, 4.13, respectively, followed by result interpretations.

As shown in Fig. 4.11, the expected value of $\overline{FOM}$ is around 0.9 (0.8958).
This suggested strongly that the segmented image contained most of the expected
hand skeletal anatomical borders of bones which is critical in computer-aided skeletal age scoring system in ossification localization and the bone age analysis as
well. Besides, it is noticeable that the dispersion of $\overline{FOM}$ is within a narrow range
between 0.8130 and 1.000. In terms of standard deviation, the data dispersion is
only 0.0502 which suggested that the precisions or consistency of detected bone
borders is very high. In other words, the segmentation framework could produce

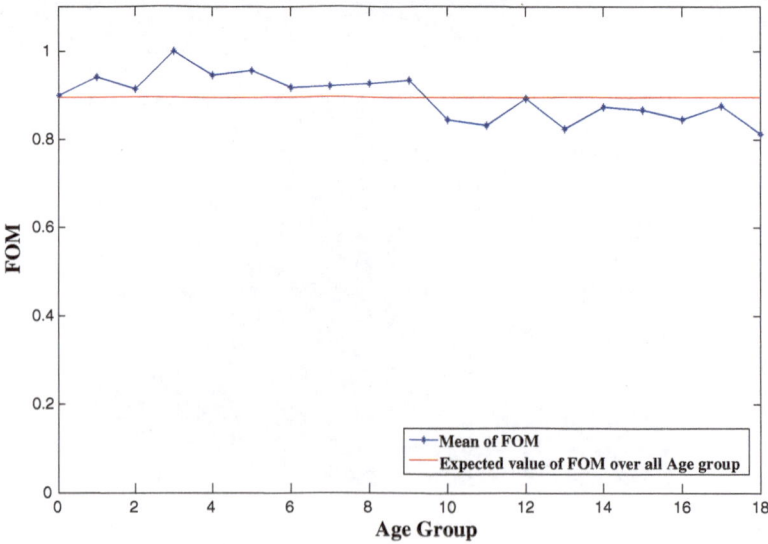

Fig. 4.11 The mean of *FOM* and the expected value

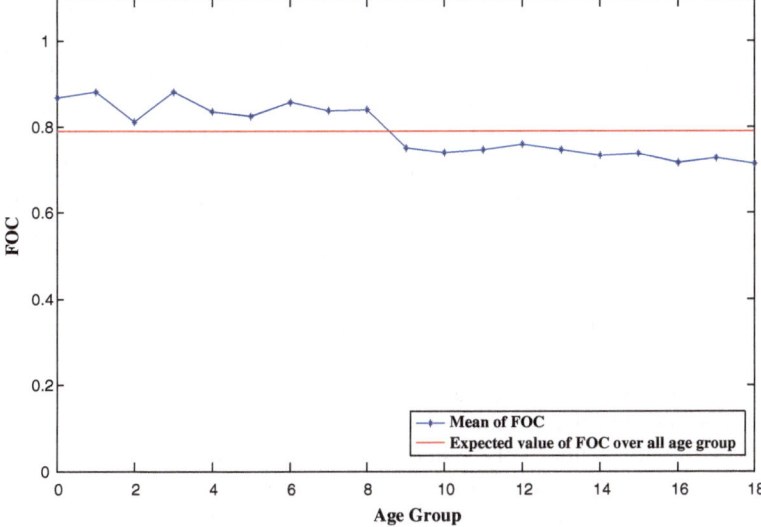

Fig. 4.12   The mean of *FOC* and the expected value

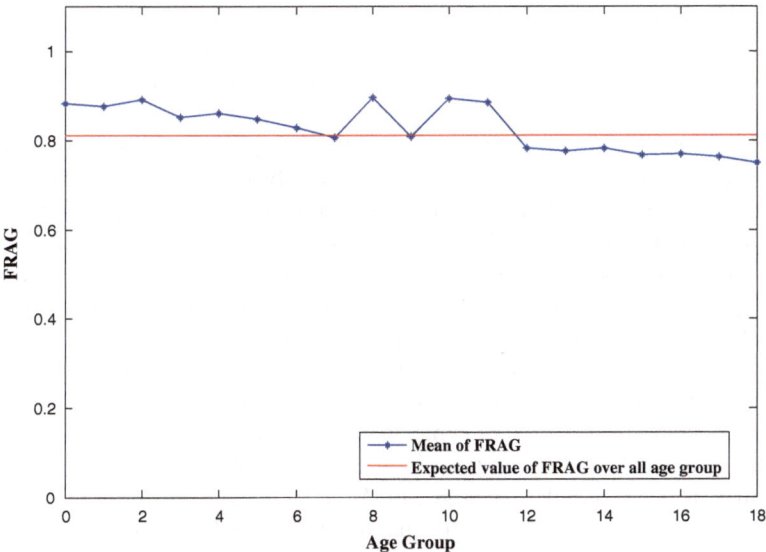

Fig. 4.13   The mean of *FRAG* and the expected value

consistent segmented hand bone edges over repetitive implementations. In comparison with radiographs before BRANEA algorithm, which has expected value of $\overline{FOM}$ as 0.5721, the BRANEA algorithm has improved the segmentation for 56.58 % to 0.8958.

As shown in Fig. 4.12, the expected value of $\overline{FOC}$ is around 0.8 (0.7897). This suggested strongly that averagely, most of the pixel intensity in actual bone have been segmented, besides it shows as well, averagely, the false labeling in ossification are in small numbers and hence it provided evidence of good segmentation. Besides, it is noticed that the dispersion of $\overline{FOC}$ is within a narrow range between 0.7130 and 0.8880. In terms of standard deviation, the data dispersion is only 0.0608 which suggested that the precisions of detected bone borders is very high. Similar to $\overline{FOM}$, the segmentation framework could produce consistent segmented hand bone regions with correct spatial location. In comparison with radiographs before BRANEA algorithm, which has expected value of $\overline{FOC}$ as 0.5179, the BRANEA algorithm has improved the segmentation for 52.48 % to 0.7897.

As shown in Fig. 4.13, the expected value of $\overline{FRAG}$ is around 0.8 (0.8114) which suggested that the number of absence or additional bones region residing in ossification sites is not likely, otherwise the value would have fluctuated below 0.5. Besides, it is noticed that the dispersion of $\overline{FRAG}$ is within a narrow range between 0.750 and 0.8912. In terms of standard deviation, the expected deviation of $\overline{FRAG}$ from expected value of $\overline{FRAG}$ amounts to only 0.0445 which suggested that the precisions of detected bone borders is very high. This suggested the consistency of the proposed segmentation framework in producing segmented image with almost correct number of bones or less number of wrongly segmented bones. In comparison with radiographs before BRANEA algorithm, which has expected value of $\overline{FRAG}$ as 0.6997, the BRANEA algorithm has improved the segmentation for 15.96 % to 0.8114.

It is observable that $\overline{FOC}, \overline{FOM}$ and $\overline{FRAG}$ gained better segmentation results in radiographs of patients that have chronological age below 9 years old. From the figures above, it is obvious that $\overline{FOC}, \overline{FOM}$ and $\overline{FRAG}$ gained output less than their expected values. The main reason for this slight deterioration is due to the increment of structural complexity in hand bone radiographs as the age of patients increase. However, overall, the small standard deviations of $\overline{FOC}, \overline{FOM}, \overline{FRAG}a$ in radiographs across all ages discussed in previous paragraphs suggested that the proposed segmentation framework possesses high adaptability which in turn indicates that the proposed segmentation framework is capable of adapting to different types of content in hand bone radiographs with different number of ossification sites from categories of age 0 to 17.

Besides, it is observable that FOC, FOM and $FRAG$ have strong degree of correlation from the figures above. To quantify this correlation, Pearson's product-moment coefficient, r_{XY} as followings is used to compute the correlation:

$$r_{(X)(Y)} = \frac{\sum_{i=1}^{n} \left(X_i - \overline{X} \right) \left(Y_i - \overline{Y} \right)}{(n-1) S_X S_Y} \tag{4.14}$$

where X and Y denote two random variables; S_X and S_Y denote the standard deviation of X and Y, respectively; $\overline{X}$ and $\overline{Y}$ denote the means of X and Y, respectively; n denotes the number of measurements of X and Y.

As a result, the computed Pearson correlation of $\overline{FOC}$ and $\overline{FOM}$, represented by $^r(\overline{FOC})(\overline{FOM})$, amounts to 0.806; The computed Pearson correlation of FOC and $\overline{FRAG}$, represented by $^r(\overline{FOC})(\overline{FRAG})$ amounts to 0.824; The computed Pearson correlation of $\overline{FOC}$ and $\overline{FRAG}$, represented by $^r(\overline{FOC})(\overline{FRAG})$ amounts to 0.707. It was expected that $^r(\overline{FOC})(\overline{FRAG})$ was of high correlation because both of them depend on each other; if the segmented region has many incorrect labeling, incorrect hand structures are expected to occur, and vice versa. Similarly, the high correlation in $^r(\overline{FOC})(\overline{FRAG})$ and $^r(\overline{FOC})(\overline{FRAG})$ are expected since if edges or regions were not well delineated, then it indicated that it has high likelihood to suffer errors in $\overline{FRAG}$, and vice versa.

4.4 Summary

In this chapter, the functionality of proposed MBOBHE and the modified anisotropic diffusion have been justified. Both techniques create a suitable environment for the subsequent segmentation modules by reducing the disturbance of uneven illumination, sharpening the edge and smoothing the texture; at the same time, MBOBHE is capable of enhancing the features of ossification sites to improve the performance of computerized BAA. After that, an analytical comparison of AAM segmentation framework with the proposed segmentation framework is performed. The result showed that AAM requires a large number of user-specified parameters. Furthermore, the nature of the parameters complicated the potential of being automated and hence reduces the automaticity. From the perspective of computerized BAA, high automaticity implies that the proposed segmentation framework does not require an expert to perform the segmentation after it has been finished modeling. On the contrary, the AAM requires an expert to perform the task because non-expert does not possess the knowledge of choosing the model, searching for optimized location and determine the number of iterations. This in turn indicates that the proposed segmentation framework can be operated by even non-expert, and in fact, in this book, the fully automated segmentation framework where the users need not any extra knowledge to execute has been designed of which possesses perfect repeatability. In terms of accuracy, the segmentation improvement of both the automated fuzzy quadruple division scheme and the quality assurance process in the proposed segmentation framework has been justified. Then, the proposed segmentation framework has been justified to possess high level of accuracy and low level of errors despite using much less number of parameters than AAM. The low standard deviation indicated high adaptability to radiographs containing different type of hand bones found in different age groups. The AAM, despite having the potential to gain higher accuracy of segmentation, is perceived being productively unrealistic in terms of production theory [13] because the added accuracy advantage can be perceived as a 'diminishing return' where the acquired accuracy fails to increase proportionally to additional consumption of resources or uncertainties.

References

1. Pratt WK (1972) Generalized Wiener filtering computation techniques. IEEE Trans Comput 21:636–641
2. Harwood D, Subbarao M, Hakalahti H, Davis LS (1987) A new class of edge-preserving smoothing filters. Pattern Recogn Lett 6:155–162
3. Zhang H, Fritts JE, Goldman SA (2008) Image segmentation evaluation: a survey of unsupervised methods. Comput Vis Image Underst 110:260–280
4. Thodberg HH, Kreiborg S, Juul A, Pedersen KD (2009) The BoneXpert method for automated determination of skeletal maturity. IEEE Trans Med Imaging 28:52–66
5. Niemeijer M, Van Ginneken B, Maas CA, Beek FJA, Viergever MA (2003) Assessing the skeletal age from a hand radiograph: automating the tanner-whitehouse method. In: Sonka M, Fitzpatrick JM (ed) In: Proceedings of the 2003 SPIE medical imaging. vol 5032 II, pp 1197–1205, San Diego
6. Thodberg HH, Rosholm A (2003) Application of the active shape model in a commercial medical device for bone densitometry. Image Vis Comput 21:1155–1161
7. Fenster A, Chiu B (2005) Evaluation of segmentation algorithms for medical Imaging. In: Engineering in Medicine and Biology Society, 2005. IEEE-EMBS 2005. 27th Annual international conference, Shanghai, pp 7186–7189
8. Unnikrishnan R, Pantofaru C, Hebert M (2007) Toward objective evaluation of image segmentation algorithms. IEEE Trans Pattern Anal Mach Intell 29:929–944
9. Polak M, Zhang H, Pi M (2009) An evaluation metric for image segmentation of multiple objects. Image Vis Comput 27:1223–1227
10. Zhang YJ (1996) A survey on evaluation methods for image segmentation. Pattern Recogn 29:1335–1346
11. Pratt W (2007) Digital image processing. Wiley-Interscience, New York
12. Strasters KC, Gerbrands JJ (1991) Three-dimensional image segmentation using a split, merge and group approach. Pattern Recogn Lett 12:307–325
13. Nicholson W (2008) Microeconomic theory: basic principles and extensions, 11th edn. South-Western College Pub, Stamford

Chapter 5
Conclusion and Future Works

5.1 Conclusion

The main book objective has been achieved by designing a segmentation framework that is invariant to both the limitations but at the same time exhibiting accurate and consistent in comparisons to basic segmentation approaches. As a result,
empirical and analytical experiments showed that the proposed segmentation
framework was able to achieve relatively high accuracy despite the low number
of user-specified parameters and no training sample is needed. This property of
invariance, automaticity and other desired properties of segmentation have been
considered and incorporated in the designed framework in order to make it becoming more practical in many daily applications such as the segmentation stage in
computer-aided skeletal age scoring system. This proposed framework minimizes
the labor-intensity, time consumption and high dependency on operator expertise
and subjectivity; automaticity also implies perfect repeatability and consistency
advantage.

The possible segmentation techniques and the current hand bone segmentation
techniques by analyzing the techniques applicability in automated hand bone segmentation have been critically reviewed and evaluated to achieve the first objective. The evaluation comprises of computational complexity, number and nature of
user-specified parameters, the requirement for expert interventions during execution phase, the consistency of performance, and dependency on certain features,
the effect of inherent property of hand bone radiograph on each segmentation
technique.

The eventual purpose of existing histogram equalization has been redefined to
extend the capability to solve variability problem to achieve the second objective.
This solves the variability problem across different radiographs or radiographs
of patients from different age group by proposing a holistic histogram equalization technique that considers luminance changes, the detail loss and the enhanced
contrast.

A fully automated anisotropic diffusion has been proposed, designed and
implemented to achieve the third objective. This automates conventional anisotropic diffusion that requires manual intervention to specify the diffusion strength
and to select the scale. The anisotropic diffusion is able to smooth the irregular

and uneven texture within soft-tissue region, within the cancellous bone, within the cortical bone and within the radiographic background. This smoothing process is conductive in mitigating the inferior segmentation effect resulting from non-uniformity within specific region and overlapping distribution of intensity range.

The quadruple division has been proposed and inserted into the segmentation framework to instill a certain extend of human cognitive ability by using the rule-based fuzzy inference system to achieve the forth objective. This scheme successfully searches optimized size in applying the central fundamental algorithm automatically according to the texture information of segmented region without presetting the size and number of division. This scheme realizes the requirement of the framework in terms of automaticity and adaptability.

A post-processing quality assurance scheme has been proposed and applied in the segmentation framework to achieve the fifth objectives. As the result has shown, this scheme is capable of supplying the ACR algorithm to the preprocessed hand bone radiograph with the most suitable size of block by modeling human knowledge implicitly. The process is vital in assessing the segmented image and determines whether there is a need to restore the lost detail, eliminate unwanted segmented regions or remain as it is.

Generally, the contribution of this study is that it adds substantially to the understanding of the possibility to solve a complicated segmentation problem by incorporating merely any relatively simple algorithm together with a series of customized modules containing a certain level of prior background knowledge of human. Lastly, the main contribution of the proposed technique is not merely to establish a novel segmentation framework to specifically resolve the hand bone segmentation problem but to introduce a new generic segmentation framework with the mentioned concept that has high generality to be applied and modified in different biomedical image processing applications. The segmentation framework, as a whole, fosters the insight about recognizing the inherent limitations of techniques and input information and then knowing how to extract their relational patterns via critical analysis to derive relevant measures for the purpose of rendering the optimized 'environment' and 'material' for a relatively fundamental algorithm to leverage the performance of which to a level that is comparative to sophisticated algorithms that involve unpractical constraints.

5.2 Future Works

This research has opened up new research questions in need of further investigation. Firstly, the future exploration on optimum kernel size in filtering processing or edge detection process is necessary to optimize the adaptability and automaticity on various dataset from different input sources. The kernels play a pivotal role analogous to human 'eyes' to process the local information in image. It remains a challenge to choose the optimum size of kernel that able to emulate human cognitive visual ability. Secondly, it would be interesting if the priority of multiple objectives in histogram equalization can be tuned autonomously according to desired segmentation output without any assumptions. Thirdly, future research should attempt to investigate the possibility in transforming more human intelligence, perception, cognitive ability and prior knowledge into the segmentation framework to reduce the number of modules and hence it would further improve the computational efficiency. In terms of generality, the quadruple division scheme can be improved to undertake any arbitrary shape of block instead of rectangular but yet remain practically feasible. As further refinement, a certain level of learning process should be imposed on the segmentation framework by analyzing the output or by-products of each previously accomplished segmentation task on radiograph in order to model, generalize and predict the underlying pattern for facilitating and improving the subsequent segmentation performance on similar radiographs as the number of segmented radiographs increases. Besides, further research should focus on extending the proposed segmentation framework to 3D image segmentation or any real-time application. Last but most importantly, the insight gained from the study should be implemented in more segmentation tasks in different applications of different field and in different context to further generalize it into a more general framework that can be arbitrarily adopted in more complicated segmentation problem in the future.

Index